Guidelines for
EMPLOYEE HEALTH PROMOTION PROGRAMS

Association for
Worksite Health Promotion

Written for AWHP by

William B. Baun, MS
Tenneco Health and Fitness

William L. Horton, MBA
Fitness Systems

Jean Storlie, MS, RD
JS Associates

Human Kinetics

Library of Congress Cataloging-in-Publication Data

Baun, William B.
  Guidelines for employee health promotion programs / Association
for Worksite Health Promotion : written for AWHP by William B.
Baun, William L. Horton, Jean Storlie.
      p.   cm.
  Originally cataloged under title m.e. in 1992.
  Includes bibliographical references and index.
  ISBN: 0-87322-351-9
  1. Health promotion.   2. Industrial hygiene.   I. Horton, William
L.   II. Storlie, Jean.   III. Association for Worksite Health
Promotion.   IV. Title.
  III. Title.
  RC969.H43G85   1994
  658.3'82--dc20   94-29747
                                                      CIP

As of January 1, 1993, the Association for Fitness in Business became known as the Association for Worksite Health Promotion.

ISBN: 0-87322-351-9

The photos on the following pages were provided by Fitness Systems, Los Angeles: pp. 3, 14, 22, 66, 69, and 82, courtesy of GE Capital Corporation Fitness Center, Stamford, CT; pp. 34, 35, 62, 67, 76, and 81, courtesy of GE Aircraft Engines, Health and Fitness Center, Lynn, MA; and p. 24.

All other photos were provided by Tenneco, Inc.—Matthew H. Alexander, photographer.

Developmental Editor: Christine M. Drews; **Assistant Editors:** Elizabeth Bridgett and Kari Nelson; **Copyeditor:** Wendy Nelson; **Proofreader:** Molly Bentsen; **Indexer:** Schroeder Indexing Services; **Production Director:** Ernie Noa; **Typesetters:** Sandra Meier and Kathy Boudreau-Fuoss; **Text Design:** Keith Blomberg; **Text Layout:** Tara Welsch; **Cover Design:** Jack Davis; **Interior Art:** Jim Hampton and Thomas Janowski; **Printer:** Braun-Brumfield, Inc.

Printed in the United States of America   10   9   8   7   6   5

Human Kinetics
Web site: http://www.humankinetics.com/

United States: Human Kinetics, P.O. Box 5076, Champaign, IL 61825-5076
1-800-747-4457
e-mail: humank@hkusa.com

Canada: Human Kinetics, Box 24040, Windsor, ON N8Y 4Y9
1-800-465-7301 (in Canada only)
e-mail: humank@hkcanada.com

Europe: Human Kinetics, P.O. Box IW14, Leeds LS16 6TR, United Kingdom
(44) 1132 781708
e-mail: humank@hkeurope.com

Australia: Human Kinetics, 57A Price Avenue, Lower Mitcham, South Australia 5062
(08) 277 1555
e-mail: humank@hkaustralia.com

New Zealand: Human Kinetics, P.O. Box 105-231, Auckland 1
(09) 523 3462
e-mail: humank@hknewz.com

Contents

Preface   v

Introduction   vii

Quality Standards for an Employee Health Promotion Program   xi

Phase I   Initial Planning   1

Management Commitment and Support   2
  Communication Channels   2
  Preliminary Financial Commitment   7

Needs Analysis   10
  Managers   11
  Employee Population   13
  Internal Resource Appraisal   14
  External Resource Appraisal   15

Summary   16

Phase II   Conceptual Definition   17

Philosophy and Scope   18
  Mission Statement   19
  Goals   19
  Priorities   21
  Evaluation Plan   22

Program Design   22
  Program Mix   23
  Marketing Strategy   30
  Staffing Model   32
  Facility Plan for Companies Developing Fitness Centers   36
  Facility Plan for Companies Developing Educational Programs   39
  Equipment Needs   41
  Financial Plan   45

Summary   48

Phase III   Implementation   49

Program Activation   50
  Program Calendar   50
  Enrollment and Health Screening   54
  Health/Fitness Assessments   57
  Counseling and Orientation   58
  Retention and Motivation   60

Marketing and Promotions   62
  Launching a Marketing Campaign 63
  Market Analysis and Monitoring 65
Staff Selection   66
  Qualifications and Certifications 66
  Recruitment 67
  Internship Programs 68
Operations and Administration   68
  Policy and Procedure Manual 69
  Facility and Equipment Maintenance 69
  Record Keeping 70
Summary   72

Phase IV Evaluation   73
Project Evaluation   74
  Outcome Evaluation 74
  Impact Evaluation 76
  Process Evaluation 76
  Cost-Effectiveness Analysis 77
Periodic Reviews   77
  Quality Assurance 78
  Monthly Review 80
  Quarterly and Semiannual Reviews 81
  Annual Review 81
Longitudinal Data Analysis   81
  Behavior Change 82
  Cost-Benefit Analysis 82
Results Interpretation and Communication   82
  Staff 82
  Participants 83
  Employee Committee 84
  Management 84
  Data-Base Management 84
Summary   85

Appendix A   Glossary of Practical Terms   87
Appendix B   Guidelines for Management Survey   91
Appendix C   Sample Employee Surveys   97
Appendix D   Resource List   103
Appendix E   Case Studies   107
Appendix F   Job Descriptions   113
Appendix G   Sample Table of Contents for Policy and Procedure Manual   121
References   123
Recommended Readings   127
Index   129
About the Authors   133
Corporate Sponsors   135
Twenty Years of Leadership and Growth in Worksite Health Promotion   137
Association for Worksite Health Promotion Fact Sheet   138
Membership Application   139

Preface

In the 1970s companies began to implement worksite fitness programs as a demonstration of support for their employees. These programs brought about positive changes in employee health, and they increased productivity and employee morale. Worksite fitness programs evolved into comprehensive health promotion programs encompassing multiple dimensions of employees' lives. As health care cost containment became a major issue in the 1980s—and American businesses were forced to assume a greater share of those costs—economic factors began to drive most decisions related to employee benefit programs. Cost-benefit and cost-effectiveness studies appeared in the health promotion and fitness literature, and the results of these studies were generally positive. Since then, more American companies have created health promotion programs for their employees, and worksite health promotion has been proposed as part of a solution to the nation's health care problems.

The Association for Fitness in Business (AFB; now called the Association for Worksite Health Promotion, or AWHP) receives approximately 700 to 1,000 inquiries annually from companies soliciting advice on how to initiate an employee health promotion program. To address this growing concern for corporate health promotion, AFB commenced the development of two publications in 1989: The first, this publication, provides guidelines and standards for companies to use in developing these programs, and the other focuses on the economic impact of worksite health promotion. (For more information on the cost-containment rationale for employee health promotion, readers should refer to the second publication, *Economic Impact of Worksite Health Promotion.*) *Guidelines for Employee Health Promotion Programs* outlines the process and quality standards that will result in a successful employee health promotion program. As such, it is designed as a practical tool, outlining key action steps that will guide corporations through the start-up phases of an employee health promotion initiative.

To develop these guidelines, AFB established a writing team of three experts involved in the development and management of health promotion programs. Collectively, these authors have over 40 years of experience, bring the perspectives of both in-house and contract management of employee health promotion programs, and have formal education in nutrition, exercise, and business. In addition to this writing team, AFB assembled a review board of seven authorities in health promotion. This manuscript was subjected to a two-tier review process. Initially, the writers developed a detailed outline for the book and a conceptual model for the guidelines. The outline and model were both critiqued by the reviewers, whose feedback was incorporated into the writing process. An extensive review and revision process was used to incorporate a broad perspective on the quality standards for the development of a successful program. This book does not promote the development of one program model, but provides the framework for designing a program tailored to the special needs of each worksite.

Companies of all sizes can follow the process outlined in these guidelines to design and implement programs that fit their employee populations and financial resources. Large companies will find the in-depth information on budgeting, facility design, and staffing useful in undertaking an extensive program development project. Small and medium-size companies can selectively use the detailed information in areas that are applicable to their specific goals. This book includes many tables, figures, and appendixes, which allow for quick reference to supporting detail. It has been written and designed so that both executives and program developers can benefit from the information. Executives can skim the headings, tables, figures, and bulleted points to familiarize themselves with key issues. Program developers will find the book a useful desk reference during the months of start-up and the initial year of program development. However you plan to use this book, AFB hopes that is will assist you in the design of a top-notch program that will, in turn, enhance the caliber of your company's work force.

Introduction

Many terms are used to describe the concept of health promotion or disease prevention. In 1960, Dunn first applied the term *wellness*, referring to *"the process of adapting patterns of behavior that lead to improved health and heightened life satisfaction."*[1] Installing wellness programs at the worksite involves not only the setting in which a wellness program is delivered but also the employer's active sponsorship of the wellness program. *Health promotion* is another term commonly used in reference to worksite wellness programs. In fact, the two terms are frequently used interchangeably. The role of a sponsoring organization in the wellness process is inherent in Opatz's definition of health promotion: "the systematic efforts of an organization to enhance the wellness of its members through education, behavioral change, and cultural support."[2]

Implicit in both of these definitions is the belief that each person can enhance the quality of his or her life through a continual process of lifestyle improvement and balanced living. Over the last 15 years, corporations have begun to take an active role in this aspect of their employees' lives because the quality of their lives affects the overall productivity, health, and stability of a work force. Approaches to employee health promotion vary greatly from company to company. To illustrate this point, the following table presents various types of health promotion programs found in companies of various sizes.

Percent of Worksites Offering Specific Health Promotion Activities

Type of activity	Worksite size (number of employees)			
	50-99	100-240	250-749	750+
Health risk assessment	18.4	34.0	41.8	66.2
Exercise/fitness	14.5	22.7	32.4	53.7
Back care	19.5	34.8	41.4	47.4
Smoking cessation	30.1	37.5	39.5	57.9
Weight control	8.1	13.5	22.9	48.8
Nutrition education	8.6	19.8	21.9	48.0
Stress management	14.5	32.7	37.5	60.8

Note. From *National Survey of Worksite Health Promotion Activities* (p. 10) by U.S. Department of Health and Human Services, Public Health Service, Office of Disease Prevention and Health Promotion, 1987, Silver Spring, MD: ODPHP National Health Information Center.

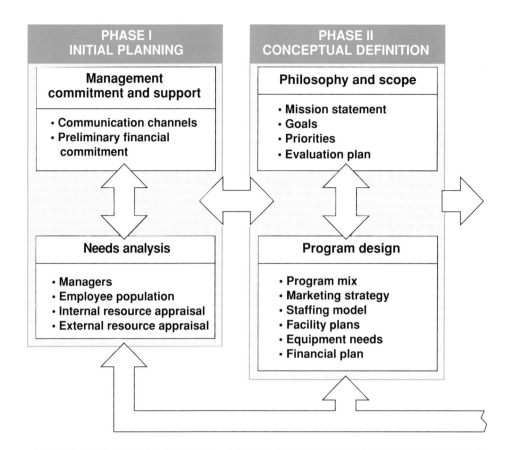

Process for the development of an employee health promotion program.

Note that in smaller size companies (50 to 99 employees), the most common type of program is smoking cessation (30.1%). This arises out of growing social pressures to establish clean air policies and the responsibility of companies to provide smoke-free air for their employees. Smoking cessation also has a well-documented cost-benefit ratio. As company size increases, a greater percentage of the companies offer many different programs. In companies with more than 750 employees, between 48% and 66% of the companies offer each specific program listed. This is because larger companies more frequently have the resources and employee base to offer a broad range of program options. However, this data should not discourage the smaller company; it is possible to realize the benefits of employee health promotion by selectively offering programs that meet the particular needs of an employee population.

Because any of these specific programs could be effective in one setting and inappropriate in another, the guidelines do not promote one program model for an employee health promotion program. Instead, the model that is used *defines a process* that guides a company through the development of a program tailored to meet its corporate philosophy, goals for health promotion, and employees' needs. The figure above illustrates the four-phase process introduced in this book. This model forms our structural framework and is referred to throughout the guidelines; each phase in the model is described in one chapter.

The first phase, Initial Planning, lays the foundation for further development. In this phase, two critical steps—obtaining manage-

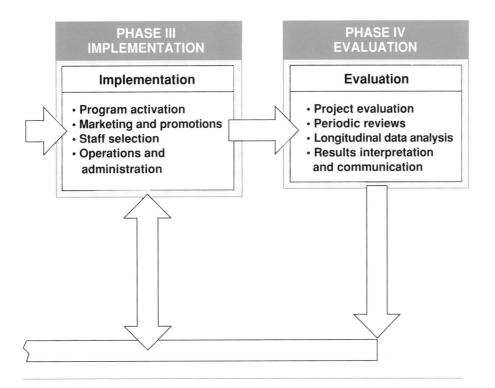

ment support for the project and determining the company's needs through needs analysis—will shape the program design (Phase II) and influence the success of early operations (Phase III). Fostering management commitment and soliciting adequate resources usually depends on some preliminary needs analysis, and vice versa; which comes first will vary from company to company. As the model suggests, these two steps may even be pursued concurrently, but it is important that both of these activities are completed before the program design is finalized and implementation begins.

During Phase II, Conceptual Definition, the corporate needs and resources that were identified in Phase I will help to structure and customize the program design. It is critical during this stage to clearly define the mission and purpose of the program and to establish specific, measurable goals that can be used later in evaluation. Program planning becomes more concrete at this stage, when decisions are made regarding the program types, time schedules, staffing, equipment and facility needs, and proposed budget.

In some settings Phase II may precede Phase I, particularly when the program is a "grassroots" initiative. For example, a program model may be proposed to management, then researched and reviewed before approval for implementation is granted. The feedback from management and results from the needs analysis usually lead to substantial revisions in the program design. Clearly these two phases are interrelated, and regardless of the starting point, they lay important groundwork for later efforts.

In Phase III, Implementation, the plans developed in the initial stages are activated. This chapter addresses the preliminary operational needs, such as hiring and orienting staff, setting up program schedules, and launching a marketing campaign. Even though operations become more refined and routine in mature programs, new challenges continually arise, and management goals must focus on how to maintain high levels of participation after the initial excitement fades. However, a discussion of all the issues associated with managing a mature program is beyond the scope of this book.

Phase IV, Evaluation, discusses various practical techniques that can be used to review program outcomes and measure successes (and failures). The appropriate approach to evaluation depends to a great extent on the available resources, corporate objectives for the health promotion program, and the scope of program design. This discussion is intended to familiarize the reader with the different forms of evaluation and how they may be applied to a worksite wellness program.

The data that is collected and analyzed in Phase IV provides important feedback for Phases I, II, and III. For example, the program results could serve as important ammunition in maintaining management's commitment to the program and obtaining approval for program expansion. Likewise, the results will assist the planning team in creating new programming ideas and revising operational systems. Evaluation should not be viewed as the final step in program implementation, but rather as another critical stage in program development. The system of feedback loops depicted in the model emphasizes the fact that development, implementation, and evaluation are ongoing activities in the evolution of an employee health promotion program.

Quality Standards for an Employee Health Promotion Program

To be successful, an employee health promotion program needs to meet a variety of standards:

1. Commitment from senior management to dedicate sufficient resources—funding, personnel time, equipment, and facilities. Ideally, the management also shows support by participating in the program.

2. A clear statement of philosophy, purpose, and goals that declares the organization's commitment to motivate and assist a significant proportion of employees to practice healthier lifestyles.

3. A process of assessing organizational and individual needs, risks, and costs.

4. Leadership from well-qualified health/fitness professionals in the program's design, implementation, and ongoing operations.

5. A program design that addresses the most significant health risks to our nation, specific risks within the employee population, and needs of the organization.

6. High-quality and convenient programs that motivate participants to achieve lasting behavior changes.

7. Effective marketing to achieve and maintain high participation rates.

8. Efficient systems for program operation and administration.

9. Procedures for evaluating program quality and outcomes.

10. A system for communicating the program results to employees, staff, and senior management.

Initial Planning

Management commitment and support
• Communication channels • Preliminary financial commitment

Needs analysis
• Managers • Employee population • Internal resource appraisal • External resource appraisal

Phase I, Initial Planning, can take 3 months to 1 year, depending on the corporate priority of the program, the enthusiasm and competence of key players, and the sophistication of the proposed program. Although the planning process may seem tedious, the time invested will pay off in the next phase, Conceptual Definition. In fact, a frequent mistake is to move forward with program implementation before sufficient planning has been completed.

The Initial Planning phase involves two stages: (a) obtaining management commitment and support, and (b) conducting needs analysis. As the figure on pages viii and ix depicts, these two stages are interdependent; the work completed in one stage promotes progress in the other. In one company the starting point may be preliminary commitment from management, followed by a needs analysis, and then allocation of resources. In another situation, the needs analysis may be performed and the results used to garner support from management. The sequence is not critical, but the steps involved in each stage are important.

Management Commitment and Support

To be successful, an employee health promotion program must have strong backing from top-level management. This is critical for securing resources for start-up and operations, conducting internal promotions, establishing a corporate culture that is health oriented, and generating optimum participation rates. This section addresses the steps involved in structuring the employee health promotion initiative to secure management's support: (a) establishing communication channels, and (b) securing a preliminary financial commitment.

Communication Channels

A variety of individuals and departments will play a role in the development and supervision of an employee health promotion program. It is important to establish a system for coordination and communication early in the planning process, even though the structure may change as the program matures. Critical to the program's success is a framework that allows for access to top management. A program sponsor (sometimes referred to as the program "champion") may emerge from one of several departments and play a crucial role in program development. As a program evolves, various individuals could become involved, from employee lay leaders to external consultants.

Communicating through a defined organizational structure will help to clarify these relationships and minimize confusion. A few of the key players and their roles, as well as possible organizational structures, will be discussed.

Access to Top Management.   Employee health promotion initiatives may originate from a number of sources; consequently, the ways in which the programs take shape and develop identities within companies will vary. In some cases, the CEO will spearhead the program start-up. This can be a desirable situation—the program initiative receives priority attention and the necessary resources are more readily obtained, and selling the concept internally is also facilitated because resistance from other sources can be more easily controlled. Later, during the promotion and implementation of the program, the CEO's direct involvement will be helpful because many employees tend to model their behavior after that of the corporate leader.

The initial idea for a worksite health promotion program could also come from the benefits manager, company nurse, medical director, or even a group of employees. When the concept is initiated from a source other than top-level management, a critical stage in implementation will be to secure management's commitment to the program. This stage can take anywhere from 2 to 8 months and involve data collection, proposal writing, and persistence.

Role of a Champion.   The initiative could falter and die at the idea stage without a program "champion," who might be the CEO or another employee. Typically, the individual who originally proposes the idea becomes the champion. Serving in this role involves more than a set of technical or management skills; the champion must possess certain personal qualities to succeed. Important characteristics of a champion are listed in Table 1.1. This key person must believe in the program and be willing to persevere when the project does not receive immediate support. An emotional, as well as professional, investment is required from a champion; because of this, it is most often a voluntary commitment. A person who is well positioned in the company, personally committed to wellness, and influential with the employees will be most effective as a program champion. It could be the CEO or another top manager, but the champion does not necessarily have to belong to this upper echelon. However, top managers need to recognize when a staff member demonstrates initiative and other essential attributes, then they need to give her or him the freedom and resources to further develop the concept. The champion also needs continuous access to top management to function effectively in this capacity. Management's role in positioning a champion involves both *identification* and *empowerment* of the right person.

Table 1.1   Characteristics of a Champion

Positioned with access to top management
Well respected by management
Visible and well liked by employees
Risk taker
Fitness/health role model
Articulate and persuasive

Technical Expert. Early in the process, the company will need the expertise of someone trained in health and fitness, either internal or external to the company, to advise on the technical aspects of program development. Depending on the scope of the proposed program, the company can hire either a full-time professional, a part-time consultant, or a firm that specializes in corporate health/fitness. During initial planning, external consultants may be best, because they generally bring experience from a variety of settings that can be valuable when the program is still being defined.

If the company expects to hire one or more full-time professionals, it may want to recruit the program director during the initial planning stage, because he or she will inherit many of the decisions made during this first phase. Moreover, a professional can assist with technical issues and provide guidance in the development process. The sections on staffing, in Phase II, Conceptual Definition (p. 32), and Phase III, Implementation (p. 66), further discuss the factors to consider in hiring health/fitness professionals.

The Association for Fitness in Business publishes an annual directory of resources to guide companies in locating services and products.[3] A company that decides to hire an external consultant for either start-up or ongoing management will find this directory useful. Table 1.2 outlines several factors to consider in hiring a consultant.

Table 1.2   Criteria for Selection of a Health/Fitness Consultant

Does the organization manifest *professionalism*?
☐ High ethical and competency standards
☐ Objectivity in analysis and recommendations
☐ Sensitive information treated with confidentiality

Does the organization have a solid *track record*?
☐ Number of projects/client list
☐ Client organizations similar in size, employee demographics, dispersion, culture
☐ Results from other programs (e.g., participation)
☐ Client references

Who will manage the project?
☐ Size of team
☐ Credentials
☐ Experience
☐ Image
☐ Enthusiasm

How will the *results* be presented?
☐ Report most suitable for top management
☐ Informal progress reviews
☐ Comparability of recommendations to national norms

How will *follow-up* be handled?
☐ Follow-up systems available
☐ Reporting mechanisms

What are the overall *capabilities* and *credibility*?
☐ Full-service capabilities or single-program focus
☐ Expertise
☐ Cost and quality

The services of an outside consultant may include any of the following activities:

- Needs analysis/feasibility study
- Program definition
- Program development/customization
- Facility design
- Equipment recommendations
- Financial analysis
- Staff training and recruiting

Employee Advisory Committee. Just as important as top management support is the active participation of employees at all levels of the company. One effective technique for eliciting employee support is to establish an employee advisory committee with representatives who serve as ambassadors for the program and spokespersons for their work groups. In structuring an employee committee, it is wise to have representation from all departments and all levels. The committee can provide advice on the program design and marketing plan and participate in special events and promotions.

A frequent mistake when establishing an employee committee is to grant decision-making authority to the employee committee. This can be disastrous because management by committee is slow and politically complicated.[4] Members of a committee are typically not professionally trained, yet are personally involved in health/fitness, so they tend to underestimate the need for professionals to supervise an effective program and reach the sedentary population.[5] Consequently, they make their recommendations and decisions based on their own personal preferences, not necessarily what is the best match for the employee group as a whole.

Despite these drawbacks, an employee committee can be an important asset to the program if it is well managed. This responsibility is a critical function of the health/fitness professional, who must possess the diplomacy and leadership to field an employee committee's ideas and channel them into workable plans. It is important to clarify the following when inviting employees to participate on the committee:

- Mission of the program
- Duties of the committee
- Meeting schedule and frequency
- Length of membership on the committee

In appointing members of an employee committee, department heads and supervisors should be asked to appoint volunteers from their work groups who would be good representatives. The ideal characteristics should be clearly communicated to the managers so they have guidelines to use in identifying potential candidates. Table 1.3 outlines factors to consider in selecting employees for the advisory committee.

Organizational Structure. The coordination of all the various individuals who may play a role in program development will be facilitated if an organizational structure is established for the program. The various reporting channels may evolve over time, but a temporary framework should be set up during the initial planning stage. Later,

Table 1.3 Factors for Selection of Employee Committee

Company representation
☐ Top management
☐ Middle management
☐ Corporate staff
☐ Line workers
☐ Marketing/sales
☐ Operations
☐ Support departments (e.g., graphics, maintenance)

Employee qualities
☐ Visible in work group
☐ Outgoing personality
☐ Enthusiastic, cheerful
☐ Communication skills
☐ Creative
☐ Cooperative
☐ Some leaders and some followers
☐ Some health/fitness role models and some nonparticipants

when the program has stabilized, a more permanent organization chart can be formulated. The appropriate structure is dictated to a great extent by internal factors, but a few considerations that are applicable to many worksite wellness programs are addressed here.

If the CEO spearheads the program, the day-to-day responsibility must be delegated because of the time involved in overseeing the details of start-up. Although the CEO's involvement is important, without support from below the program could progress very slowly. In a larger company, the responsibility is typically delegated to either the human resources or the medical department. In a smaller company, an executive secretary, administrative assistant, or another staff member who works closely with the CEO could coordinate the program. If a program champion has been identified, the responsibility for program management can be delegated to her or him.

Figure 1.1 (pp. 8-9) illustrates a few organizational structures that could be adopted for a worksite wellness program. The first structure on the chart depicts a large corporate structure, but this scenario could occur in a company of any size. A talented champion has emerged, has been partially released from other responsibilities, and is charged with spearheading the project. The champion, in turn, assembles an employee committee with broad employee representation and hires a number of external consultants (vendors) to provide technical support. In this case, the champion is responsible for writing proposals, preparing status reports, and communicating with management.

In the second structure, the responsibility for the program's development and implementation has been delegated to the human resources department. Once again, an employee committee is established to provide a system for employee input. One significant difference in this structure is that an external provider organization has been hired for program management. The external organization provides the staff, program materials, and technical expertise needed for program delivery. The responsibility for overseeing the provider's contract, coordinating input from the employee committee, and communicating with management is channeled through the VP of human resources. Although this approach may seem to be more applicable to a larger company, any company could decide to rely on an external provider. The important issue is to establish a clear reporting channel for the provider through a staff member who has the capabilities and authority to negotiate contracts and to supervise the quality of work performed by a vendor.

In the third scenario, the company hires a health/fitness professional who is responsible for all aspects of program development and implementation. This is more likely to occur in large companies that plan to have an extensive employee health promotion program. In this case, the health/fitness professional would assemble and coordinate the employee committee and hire external providers as needed. In Figure 1.1 the health/fitness professional is shown with a reporting channel through the VP of human resources; however, this is not always the case. It is possible to have the health/fitness professional report to the medical department, to the training and development department, or even directly to the CEO.

The final structure presented in Figure 1.1 illustrates a situation where the CEO is the program champion and has delegated the details of program coordination to an executive secretary or administrative assistant. This assistant can schedule the committee meetings, circulate materials, coordinate the promotions, and handle paperwork associated with vendor contracts, but the CEO will need to stay involved in decision making. It may be a good idea, in these cases, to appoint an energetic manager to chair the employee committee so there is additional management talent involved. It is always risky to place a great deal of responsibility for program coordination on a staff member who does not have management experience or stature in the company.

In summary, the appropriate structure will need to be determined in the context of each company, depending on the company size, work loads of key individuals, personalities involved, and availability of external resources. In any structure, access to top management is imperative. This is important even if the reporting channel differs from the program coordinator's official position in the company. In addition, the structure should allow for input from employees and access to technical expertise through hiring external providers or health/fitness professionals. The staff member who is responsible for the program may need to be partially released from other duties during the months of start-up. It is important to be realistic in delegating the work load to existing staff members, or the people who become involved in the project may feel overwhelmed, develop a negative attitude toward the program, and transfer this attitude to other employees.

Preliminary Financial Commitment

Although a final budget will be determined as a result of program planning, it is helpful during the initial planning phase to estimate the level of funding that might be feasible. Management needs to make a preliminary financial commitment to investigate the concept and to define the program, with the understanding that a final budget will be presented at a later date.

To accomplish the goals of initial planning, a company can expect to dedicate 50 to 200 hours of the champion's (or program coordinator's)

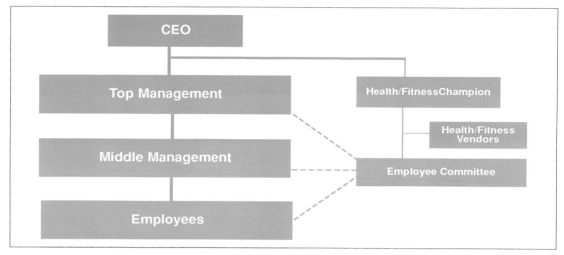

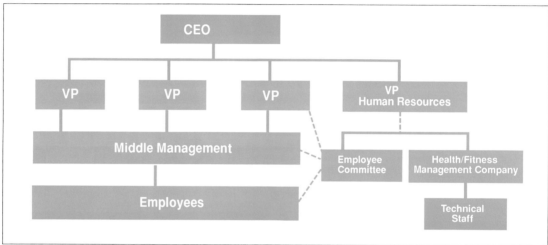

Figure 1.1 Possible organizational structures. ———— Reporting Channel – – – Committee Representation

time and $2,500 to $25,000 in consulting fees. The consulting fees will vary depending on the scope of work performed and the nature of the proposed program. Obviously, the bigger the project, the more expertise that will be required to advise on the start-up. For example, a program design that involves a series of awareness-building and educational activities will not require the capital investment of constructing a fitness facility. In a smaller program, a health promotion consultant can advise on the development of an employee committee, design of the needs analysis process, selection of research instruments, data analysis, and the interpretation of results. This work will cost from $2,500 to $5,000; however, in some companies these steps are managed internally without the assistance of a consultant. If a fitness center is being considered, an expert in facility design, construction, and management should be hired to conduct a feasibility study and develop financial projections for the facility (e.g., construction and operating budgets). It is extremely rare that a company has the expertise to internally perform this work, and a company should be prepared to spend $10,000 to $25,000 in consulting fees.

The operation of an employee health promotion program requires financial resources in the form of personnel time, equipment and ma-

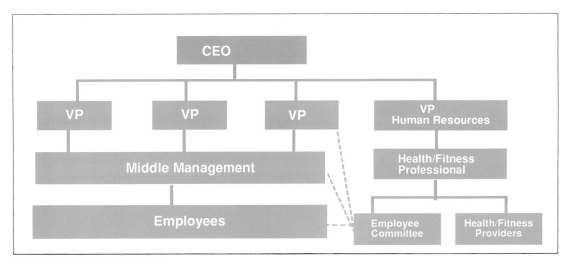

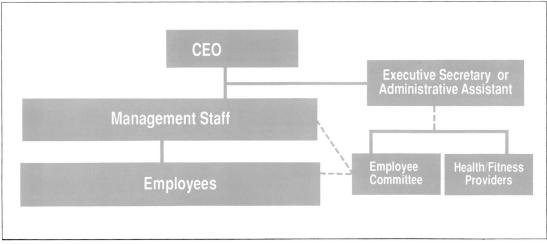

terial purchases, consultant fees, and possibly facility construction or renovation. The annual budget could be as low as $5 to $50 per employee for program materials in an awareness- or education-based program or as high as $300 to $600 per employee for a fully staffed, on-site fitness center. Table 1.4 presents cost data on program types collected by three fitness management companies. Within each program category, there is a wide range of annual costs per participant, due to the variability of program quality and delivery methods available. For example, "health assessment" could range in meaning from a health-risk questionnaire to a complete physical exam by a physician. The lifestyle-oriented programs—smoking cessation, weight control, nutrition, stress management, and preventive health—can be delivered as single lectures, full- or half-day workshops, or courses that meet weekly for several weeks. All of these factors affect the program quality and results, as well as the budget. The section on program mix in Phase II, Conceptual Definition (p. 23), addresses the qualitative issues affecting program choices in greater depth. The data in Table 1.4 may assist program planners in projecting participation rates and per-capita costs of participants, but the final operating

Table 1.4   Participation Rates and Costs of Educational Programs

Program component	Participation rate[a]	Cost per participant
Health assessment	60-80% of all employees	$ 10-50
Smoking cessation	5-10% of smokers[b]	$100-250
Weight control	10-15% of all employees	$ 50-250
Nutrition	5-10% of all employees	$ 50-250
Stress management	10-15% of all employees	$ 5-500[c]
On-site fitness center	20-50% of on-site employees	$300-600[d]

Note. Proprietary data prepared by Fitness Systems, Health Enhancement Systems, Inc., and Baxter Healthcare Corporation, 1989. Data compiled reflects the collective experience of these three companies.

[a]Assumes program is well publicized, convenient, and paid for by employer—entirely for the health assessment and essentially all for the intervention courses. [b]Typically 10% to 15% of white-collar and 25 to 30% of blue-collar workers are smokers. [c]Ranges from a 1-day workshop to a comprehensive 10-week behavior modification course taught by clinical psychologists. [d]Total net annual cost, including depreciation of capital equipment.

budget will depend on the overall program design. These financial considerations are also further discussed in Phase II (p. 45).

Employee benefits can be structured to include a per-capita allocation to health promotion. Many companies are moving toward a cafeteria-style benefits plan, where employees can select from a menu of benefit options. Health promotion could be included in the list of benefit options. Some companies offer health promotion as a copay program, where the company pays a percentage of the program fees and the participating employees pay the balance. Obviously, construction of a fitness center involves a significant capital investment and is often easier to justify during a facility relocation or renovation project. Most importantly, the company needs to approach employee health promotion as a long-term investment and be prepared to make this commitment if they expect to see a return.

Needs Analysis

The purpose of the needs analysis process is to gather internal and external data that will assist in formulating plans and proposals. This stage is important for further development because the facts gathered will help document need, customize program design, and build support. Table 1.5 outlines sources of data by impact areas to assist in planning a needs analysis project.[6] An exhaustive needs analysis should involve systematic data collection from managers and employees, and an appraisal of internal resources, as well as an assessment of external resources and other program models. This section describes the steps involved in completing a detailed needs analysis; however, each company can determine the scope of analysis necessary and adapt this process to suit its situation.

Table 1.5 Sources of Data for Needs Analysis by Impact Area

Impact area	Source of data
I. Employee needs/interests[a]	
A. Health risks	Health risk profile
B. Health habits	Needs/interest survey
C. Program interests	Needs/interest survey
D. Work partners	Needs/interest survey
E. Facility usage patterns	Needs/interest survey
II. Health benefit	
A. Medical care costs	Personnel records
B. Type of medical claims	Employee health services
C. Worker's compensation claims	Personnel records Employee health services
D. Health crises	Employee health services Anecdotes from managers
E. Life insurance costs	Personnel records
F. Other insurance costs	Personnel records
III. Productivity	
A. Morale	Employee opinion survey
B. Turnover	Personnel turnover reports Interviews with employment representatives
C. Recruiting success	Interviews with employment representatives
D. Absenteeism	Personnel records
E. Physical and emotional disabilities	Personnel records Employee health service records
F. Desire to work	Employee opinion survey

Note. From *Design of Workplace Health Promotion Programs* (p. 27) by M.P. O'Donnell, 1986, Royal Oak, MI: American Journal of Health Promotion. Adapted by permission.

[a]This section was added by the authors.

Managers

Top and middle managers' viewpoints will be important to the overall program structure and approach. While their support and participation facilitates implementation, a lack of managerial support can undermine the implementation process. Management's concerns about the program's impact on productivity, employee morale, and other operational issues should be factored into the program structure. Surveying the managers involves them early in the program and can substantially eliminate barriers later in the process. Analysis of managers may be accomplished with one-on-one interviews, a structured focus group (i.e., a group interview led by a trained market researcher), a written survey, or informal meetings. Face-to-face interviews with each manager are the most thorough approach, but they

can be expensive due to the time involved. A focus group or informal meeting is a more efficient approach, but group dynamics often bias the results. Written surveys provide an objective and consistent form of feedback; however, depending on the procedure used to administer the survey, the response rate could be low. It may be best to combine a few research techniques to minimize the problems associated with a single approach.

Managers' opinions should be handled confidentially—ideally by an external third party—otherwise managers may be reluctant to honestly express their views. Appendix B outlines a sample interview process, which could be adapted to a focus group or written format. When you are conducting an interview, it is helpful to begin with a brief description of the project in hypothetical terms. Regardless of the methodology used, the following issues should be addressed when analyzing the managers:

- Wellness philosophy and attitudes
- Goals and objectives
- Personal health/fitness habits

Wellness Philosophy and Attitudes.   It is important to determine the extent to which managers believe in the benefits of employee health promotion. If a CEO is a wellness enthusiast, but his or her beliefs are not shared by the management, there may be passive resistance to the program. Consequently, early implementation activities may need to target this group to convert them to program supporters. For example, a management retreat at a health resort, executive physicals, or a series of dynamic health seminars could prime the management team, encouraging them to embrace a wellness philosophy. On the other hand, if managers already believe in the value of employee health promotion, they can be used as a conduit for promotion and implementation.

It is helpful to know whether the managers will actively support their employees' participation in a wellness program and whether they believe that the program will benefit the company. The management survey should attempt to elicit concerns about work flow disruptions and employee-management conflicts, but at the same time it should provide management with an opportunity to participate in the development process.

Goals and Objectives.   During Phase II, Conceptual Definition, specific goals and objectives will be written. Determining the expectations of managers facilitates the development and approval of goals or objectives. These goals should specify expected program outcomes of a wellness program, such as smoking cessation, weight control, risk factor reduction, or increased fitness levels. Managers can be asked to rate their perception of the importance of specific goals and objectives.

Personal Health/Fitness Habits.   It is also helpful to assess the degree to which managers currently practice healthy lifestyles. The management team is often a role model for the company, so it is important to know what habits they exhibit. If the managers have poor health habits, it may be worthwhile to pursue program initiatives that promote healthy habits within this group. (This process will be much easier if the CEO is an active participant.) If a fitness center is being considered, it will be helpful to know whether the managers feel comfortable exercising in front of employees.

Employee Population

An employee wellness program must be based on the special needs and interests of specific employee groups who will be served by the program. It is impossible to tailor a program to their needs without first considering the following characteristics of the employees:

- Demographic profile
- Health risk profile
- Fitness/health promotion interests and habits
- Medical care cost data

Collection of data from personnel files, insurance records, and an employee survey is an important step in the needs analysis process.

Demographic Profile. Prior to conducting an employee survey, the personnel files, if accessible, should be reviewed to determine the demographic profile of the employee population. However, personnel policies that safeguard employee confidentiality often limit access to personnel files. If available, the data can be analyzed by employee subgroups, such as shifts, job classifications, or worksite locations, to provide a picture of the employee population as a whole. This information can also be included as part of a survey instrument. Useful demographic information to gather includes the following:

- Age, gender, and ethnicity
- Educational level
- Job classification and income
- Work hours, worksite location, and transportation patterns
- Sick leave and absenteeism patterns
- Health care utilization patterns
- Family status and size

Health Risk Profile. Some companies use a health risk profile or other screening activity to assess risk factors as part of a needs analysis. This approach includes promotion and intervention efforts in the initial planning stage and can be effective to build interest, involvement, and visibility for the program at early stages. One caution, however, is that once risks have been identified, a company should plan to follow up with risk reduction programs. When a company fails to follow through, employees have been made aware of their risks but will not know what to do about them, which often causes fears and uncertainties.

There are many validated health risk profile instruments available. AFB's annual directory includes a section on computerized health assessment tools.[3] A company should select the instrument and delivery method appropriate for its setting. This process can be facilitated by a consultant, but factors to consider in selecting an instrument include these:

- How is the questionnaire designed?
- How is the data entered and analyzed?
- How do the employees receive their results?
- Does the company receive a composite report?
- How easy is it to understand the results?
- What is the validity of the instrument and core data base?
- What is the cost of the instrument?

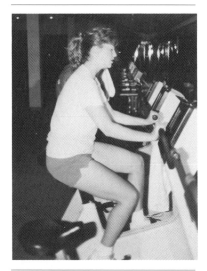

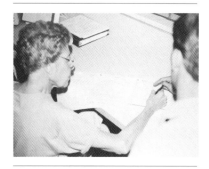

Fitness/Health Promotion Interests and Habits. When the program design stage begins, there will be a wish list of program offerings that must be prioritized. It will be helpful to know what programs the employees would like to see offered, as well as their current involvement in other programs that would compete for their time. For example, if a fitness center is being proposed, and the questionnaire reveals that 30% to 40% of the employees already belong to health clubs and are satisfied with the service, it may force management to reconsider the appropriateness of this investment and take a different approach to the employee health promotion program. Data on the employees' current health/fitness interests, attitudes, values, beliefs, and habits is most easily captured in an employee survey.

Medical Care Cost Data. If one of the goals for installing an employee health promotion program is to track and document health care cost fluctuations, it is critical to review medical claims data, workers' compensation claims, and absenteeism patterns. This can be a very time-consuming and expensive undertaking, but it will be worthwhile if the program intensity is sufficient to expect these results. It may also be helpful to engage the assistance of a utilization review or actuarial team to conduct this part of the needs assessment and evaluation process.

Employee Survey. A company-wide employee survey to assess the needs and interests of employees is crucial. Before administering an employee survey, it is extremely important that the management has made a commitment to respond in some way to the needs that are identified, or the survey could falsely raise expectations among the employees and cause dissidence. Many consultants have questionnaires and data bases to analyze the results of an employee survey and can make recommendations based on comparisons to other employee groups. An employee survey can be very extensive or relatively brief, and depends upon the scope of the proposed program. Appendix C presents two sample employee questionnaires.

Retirees and spouses may be part of the population served by the wellness program, because they typically are recipients of health benefits. If these groups are eligible for participation in a wellness program, they should be included in the survey process. Survey respondents need to be assured that the data will be handled confidentially and that it will be summarized in group reports rather than individually.

Internal Resource Appraisal

To complete the internal assessment, it will be necessary to appraise the resources available for an employee health promotion program. This will be important in determining the scope of the program, as well as incorporating existing resources into the implementation strategy. Factors to consider include facilities and equipment and staff resources.

Facilities and Equipment. If the program will be implemented in existing facilities, it will be necessary to tour the buildings and grounds to assess the possibilities and limitations. When exercise and fitness programs are being planned, this becomes a critical step in program planning and site selection. Indoor and outdoor walking trails can be planned in settings where more elaborate fitness facilities are not feasible. The practicality and safety of this low-cost option can be evaluated during a facility assessment.

When a program involves mainly educational activities, the primary facility requirements are meeting rooms of various sizes. The accessibility of meeting and conference rooms is an important consideration, because convenience is a driving force in participation and retention rates. Nutrition education programs can be implemented in the employee cafeteria, if one exists. Assessing the layout and overall environment will help determine the approach to a cafeteria nutrition education program.

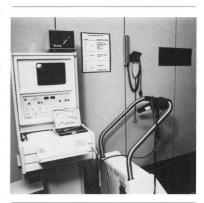

Educational equipment, such as slide and overhead projectors, bulletin boards, and flip charts, is useful in implementing health promotion activities. In many cases, this is the only equipment necessary. If not available internally, this equipment can be rented. Medical equipment, like blood pressure cuffs and weight scales, to conduct health assessments will be available in companies that have medical departments. If the company has been involved in fitness and recreational leagues or tournaments, there may even be some exercise equipment available.

Staff Resources. In addition to any health/fitness professionals who may be employed by the company, the expertise and availability of other staff should be considered when appraising internal resources. In companies that have a medical department, nurses and physicians may be able to assist in making technical decisions or even spearhead the initiative. Members of a wellness committee can sometimes contribute to the project. Management expertise will be important to shape, structure, and lead the process. Of critical importance are the other time commitments of employees involved in the program development. Reporting the program through individuals who are pressured with other priority projects will stifle progress, slowing momentum and frustrating others who are enthused and supportive. In these cases, the company should redistribute work loads so the project receives the required attention, or consider hiring an external provider.

External Resource Appraisal

Over the last 10 years, an abundance of wellness resources has been developed and marketed. Although the pioneers in corporate wellness had to develop their own programs, now a company can purchase services and products to supplement its internal capabilities, or even hire a contract management firm to install a turnkey system (i.e., implement a developed program, provide the staff and materials, and oversee the start-up and operations). It will be useful to review both local and national resources to determine the best approach. Locally, many hospitals and community organizations will provide staff to conduct on-site screening and educational programs. Universities and colleges also have experts who will consult and even deliver programs to corporations. In most large cities, there are local consultants who will assist in program and facility design, implementation, and management. In addition to AFB's annual directory of resources, there are a few other product catalogues published annually.[3] Appendix D includes a list of provider directories for health/fitness products and services. These directories can be helpful when collecting information and evaluating both internal and external options. Many purchasing decisions in the industry are made through the information obtained at trade shows and through word-of-mouth referrals.

One valuable exercise during the initial planning phase is to evaluate existing program models used in other companies. In evaluating other programs, consider the company size, demographics, wellness budget, corporate philosophy or mission, and purpose of the wellness program. Every company will have unique considerations, but similarities exist, and many companies will be proud to share their experiences and successes. Through this approach, a company can learn from others, borrowing ideas and preventing mistakes.

Summary

Phase I, Initial Planning, involves two critical steps: obtaining management support for the project and assessing the company's needs. The two steps are interdependent and can be pursued in any sequence that is appropriate for the setting. Management's commitment to the program will be necessary before resources can be allocated and structures established for project management. Various people play a role in the development of an employee program, and their efforts need to be coordinated through a defined system that allows for access to top management. The preliminary financial investment required for the research and planning process usually needs management approval. The needs analysis process provides useful data that can help document need, customize the program design, and build support. This analysis should target top and middle management, the employee population, and other program participants (e.g., families, retirees). An appraisal of internal and external resources helps determine program possibilities, potential limitations, and the investment required. The initial planning phase will shape the program design (Phase II) and influence the success of early operations (Phase III).

Conceptual Definition

Philosophy and scope

- Mission statement
- Goals
- Priorities
- Evaluation plan

Program design

- Program mix
- Marketing strategy
- Staffing model
- Facility plans
- Equipment needs
- Financial plan

It is possible to pursue some of the activities in Phase I and Phase II concurrently; indeed, many steps will overlap. As the figure on pages viii and ix depicts, these two phases are interdependent. During Conceptual Definition, it may be necessary to conduct additional research or secure management's final approval. Likewise, it may be necessary to develop a preliminary concept before management will commit to the project.

As the project moves into Conceptual Definition, the program plan becomes more concrete and tangible, which lays the foundation for implementation and evaluation. This is an exciting stage because it is possible to start envisioning the program details. During Phase II, specific program plans are written to define the following: (a) philosophy and scope, and (b) program design.

Philosophy and Scope

Before proceeding too far with program design, it is important to establish its philosophical framework and to determine the overall scope of the project. The philosophy should embrace management's beliefs, commitment, and expectations for employee wellness. Therefore, the wellness philosophy will be unique to each corporate setting. Wellness is a very broad and often vague concept, and there are many ways to define it.[7] A loosely defined wellness philosophy can lead to confusion and may result in a diffusion of resources. The program might appear scattered and unfocused, and it will be difficult to measure success.

Defining a philosophy, therefore, provides a clear vision for the program. Later, when it is time to establish priorities for the program components, the philosophy of the program helps to define its scope. The following factors can be considered when defining the program philosophy:

- Purpose (e.g., public relations, recruiting, morale, retention, cost containment)
- Goals (i.e., what level and scope of impact is expected)
- Wellness model to be used or created
- Depth and breadth of the program
- Image of the program and company
- Internal versus external development and operation

When defining the philosophy and establishing the scope of the program, the wellness team should write a mission statement, establish short- and long-term goals, determine program priorities, and outline an evaluation plan. The writing of this conceptual overview is best accomplished in a small work group of three to five individuals, which might be a subcommittee of the employee advisory committee led by a health/fitness professional. Once reviewed and revised by the entire employee committee, the conceptual overview should be approved by the top management group before the committee proceeds with further definition of the program.

Mission Statement

In developing a mission statement, the writing team should review the results of the needs analysis, particularly the management philosophy, beliefs, goals, and corporate culture.[8] The management group will need to embrace and promote the mission, so their input is critical. Because employees' interests will also shape the wellness program, the results of the employee survey should be interpreted and blended into the mission statement. Other program models may serve as reference points, so a review of other mission statements may assist the writing team. It is critical for a mission statement to be realistic; a mission statement that is too lofty can build false expectations. Therefore, the writing team needs to engage in frank discussion about the financial resources the company can reasonably invest in wellness.

The mission statement should include a statement of the management's vision for the program, the scope of program offerings, the purpose for the program, quality standards, and the employee group(s) eligible for the program. Figure 2.1 presents a sample mission statement.

Goals

Within the general framework of the company's mission for employee health promotion, it is important to establish goals that are clearly understood and widely shared, especially within top management. Each goal should include a set of objectives that are measurable within the evaluation limitations of the organization. The goals should also be realistic.[6] When developing goals, it is helpful to use the results of the management interviews (p. 11) to guide the process. Any medical data gathered in the needs analysis, as well as the employee interest survey, will also help in formulating goals. Obviously, the program

---

In keeping with the company's mission statement of providing a challenging and rewarding work environment for employees, we are committed to providing wellness opportunities for all employees.

The major goal of the program will be to provide employees a work environment supportive of positive health and fitness practices. The program will focus on helping employees and their families reduce lifestyle risk factors and become better health care consumers.

A second goal will be to evaluate the effectiveness of the program and its design, using appropriate analysis techniques.

Figure 2.1 Sample mission statement. *Note.* From Tenneco Remote Site "Pipeline to Health Program" mission statement.

should aim to reflect and reinforce the overall philosophy and image of the organization.

There is a wide variety of goals that companies aspire to achieve through health and fitness programs, ranging from enhanced morale and company image to medical care cost containment. Once again, these goals must be realistic with respect to the resources allocated. Programs will not rectify significant corporate problems, such as absenteeism and health care cost reduction, unless they are of a sufficient depth and breadth to instill behavior changes in a significant percentage of the employee population (see p. 23 for further discussion of depth and breadth). Achieving most of these outcomes requires a long-term investment.

O'Donnell has developed a framework for identifying the likelihood of achieving certain organizational goals based on the level of intervention.[6] Table 2.1 presents an adaptation of O'Donnell's model, which has useful applications in establishing realistic goals and designing a program that is likely to reach its desired goals. His original

Table 2.1 Relationship Between Levels of Programs and Impact on Organizational Goals

Organizational goals	Level I: Communication and awareness	Level II[a]: Screening and assessment	Level III: Education and lifestyle	Level IV: Behavior change support systems
Image enhancement				
Relevance to product-related image[b]	1	1	2	4
Relationships with suppliers, regulators	2	2	3	3
General visibility	1	1	2	4
Recruiting	2	2	2	4
Productivity improvement				
Morale	3	3	3	4
Turnover	2	2	2	4
Absenteeism	2	2	2	4
Tardiness	1	1	2	2
Physical and emotional ability to work	1	2	2	3
Desire to work	2	2	2	3
Health-related				
Health crises and special risks reduced	1	2	2	4
Health conditions and practices improved	1	2	2	4
Health care costs reduced	1	1	2	3-4
Disability costs reduced	1	1	2	2-3
Worker's compensation costs reduced	1	1	2	2-3
Life insurance reduced	1	1	2	2-3
Other insurance reduced	1	1	2	2-3

Note. 1 = impact unlikely; 2 = impact possible; 3 = impact probable; 4 = impact highly probable. From *Design of Workplace Health Promotion Programs* (p. 22) by M.P. O'Donnell, 1986, Royal Oak, MI: American Journal of Health Promotion. Adapted by permission.

[a]Column on screening and assessment was added by authors. O'Donnell's original model considers screening and assessment programs part of awareness programs. [b]Health care organizations (e.g., hospitals, insurance companies, hospital suppliers) find that health promotion has a positive impact on their public image because of the relationship to their product.

model has been modified to include an additional level of intervention, screening and assessment, to match the classifications for programming levels used in this publication (see Table 2.2). O'Donnell categorizes screening and assessment programs as awareness programs, because he feels that screening and assessment only serve to make people aware of their health status but have limited impact on behavior change; however, the authors of this book believe that screening and assessment take the awareness process one step farther than communication programs. These issues are discussed further in the next stage, Program Design (p. 22).

Most companies would like to try to achieve all of the goals listed in Table 2.1 but do not have the resources to implement a program at the requisite level of intensity and comprehensiveness. Therefore, it is important to establish priorities, define the scope of the program, and devote sufficient resources to the accomplishment of a specific set of goals over a certain period of time.

Priorities

At the early stages of planning, it is worthwhile to brainstorm a wish list of program possibilities before determining the scope of the program. This is an exciting stage—the planning team can be very imaginative, because they do not have to be limited by practicalities. It is an excellent activity for an employee committee—group dynamics can spark creative thinking and embellishment. This process can help to instill a sense of ownership in the employee committee. It is also important, however, to have this session led by a skilled health/fitness facilitator to prevent a scattered and unproductive discussion.

At some point, the planning team needs to define the scope of the program and undertake a manageable set of activities. Setting program priorities helps in making resource allocations and establishing a timeline for phasing in various program components. These priorities should reflect the overall philosophy and purpose of the program. In defining the scope of the program, management needs to answer the following questions:

- What employee classifications will be eligible for the program (e.g., management, line workers, employees at satellite locations, retirees)?
- Will spouses and dependents be eligible to participate?
- Which program options will best accomplish the short- and long-term goals?
- How will the introduction of one set of program offerings lead to the next phase of programming or goal achievement?
- What will be the core program offerings?
- What supplemental programs will be offered? When will they be added?
- Will these programs be sponsored by the company, or will a co-payment plan be offered?

Participation rates will impact costs, so these estimates will also assist program planners in defining the scope of the program. It is possible to estimate the number of employees who might initially participate in the wellness program from the employee survey responses. The participation rates from other corporations with similar demographics can also be useful in predicting participation levels. Industry averages for participation rates in various types of health education programs are presented in Table 1.4 (see p. 10). In predicting participation rates, keep in mind that participation in worksite wellness programs tends to increase with the socioeconomic level of the employees (i.e., as indicated by job classification and education level) and also with a more centralized work force. Off-site delivery options that require driving tend to reduce participation.

Evaluation Plan

The conceptual overview should also identify the key variables that will be used in evaluating the program, as well as the methods for evaluation. Clearly defined goals and objectives are the cornerstone of the evaluation plan, because program outcomes need to be measured against the goals. The evaluation plan should include expected outcomes, on a short- and long-term basis. In the final chapter of this book, Phase IV, Evaluation, the components of evaluation are described in detail (p. 73). The evaluation plan should cover all these components and provide a strategy for involving the entire staff in the process. The time frame for data collection, analysis, and review should also be included in the plan.

In developing an evaluation plan, it is helpful to be aware of the problems inherent in evaluating the benefits of wellness programs. First of all, many wellness outcomes are intangible (e.g., employee morale, company image) and difficult to quantify (e.g., productivity). Secondly, the most concrete goals take a long time to demonstrate. For example, it may take 3 to 10 years to realize health care cost reductions as a result of employee health promotion. Although a segment of the employee population will probably experience health improvements that will affect their jobs through reduced absenteeism and turnover, it could take as long as 10 years before these changes have occurred in a large enough percentage of the employee population to affect the bottom line. Further, it is labor intensive and costly to track data and to document results consistently. Finally, even when the data is available, it is difficult to apply the rigors of sound research methodology in the corporate setting. For example, without using a control group it is not possible to attribute positive changes solely to a health promotion program as opposed to other causes.

Despite the obstacles associated with evaluating the benefits of health promotion, it is conceivable for every program offering to measure participation rates, attrition, postprogram behavior changes (e.g., smoking cessation, weight loss, fitness improvements), and employee satisfaction with the program (e.g., instructor quality, program materials, convenience, cost).

The evaluation plan should also include a process for using the data from evaluation to improve the program and stimulate program growth. Staff meetings, focus groups, task forces, and employee committees can use evaluation data to make better decisions affecting the worth and growth of the program. Therefore, a central element of an effective evaluation plan should be a diagram for how evaluation information will flow back to the different groups.

Program Design

During the program design stage, plans will become more detailed. If the previous stages have been successfully completed, it is relatively easy to develop the program specifics. For many people, this stage is the most rewarding because the process is concrete; the program moves from an idea to an operational plan. If the groundwork has not been laid in Phase I, Initial Planning, the program design and implementation can be fraught with difficulty, because the commitment might still be tentative and the needs analysis may be weak. Similarly, it will be harder to define program components if the philosophy and scope have not been clearly defined.

Program design involves more than selecting a set of program options; the design needs to address the marketing strategy, the staffing model, the facility plan, equipment needs, and the financial impact, because all of these factors are interrelated. The program mix drives many of these decisions; however, the budget also shapes staffing, facility, equipment, and, in turn, program choices. This section addresses how these program-related decisions can be adapted to various worksites.

Program Mix

The program mix is the total of all program opportunities offered to employees in a health promotion program, and it can be defined by its breadth and depth. Figure 2.2 depicts this concept. *Program breadth* refers to the program's comprehensiveness, and *program depth* refers to the number of different options offered within each area the program covers.

A comprehensive program could cover a wide range of wellness dimensions (e.g., physical fitness, emotional well-being, intellectual development, social adjustment), but these program offerings may be limited in scope and number. Such a program would have a broad/shallow program mix. On the other hand, a program may focus on making an impact in one aspect of wellness (e.g., smoking cessation) but offer many activities and incentives to promote this lifestyle change; this program would have a narrow/deep mix. Successful large health-promotion programs that have existed for several years usually have a comprehensive program that is both broad and deep. Smaller and younger programs usually have depth in one core area, supplemented by other program options.

B R E A D T H

		Narrow	Broad
D E P T H	**Shallow**	Few programs lines, and only one option in each line.	Several different program lines, and only one option in each line.
	Deep	Few program lines, and many options in each line.	Several different program lines, and many options in each line.

Figure 2.2 Breadth and depth of program mix. *Note.* From *Marketing* by Robert D. Hisrich. Copyright © 1990 by Barron's Educational Series, Inc. Reprinted by permission of Barron's Educational Series, Inc., Hauppauge, NY.

When resources are limited, a program must sacrifice either breadth or depth, or quality will suffer as the program becomes scattered and diffused. Depth is needed to reach those employees who are most resistant to change, whereas program breadth provides a sense of egalitarianism (e.g., employees who do not smoke will not benefit from a wellness program that focuses primarily on smoking cessation). A broad program may, on the surface, appear to have a bigger payoff by offering enough program options to meet everyone's needs, but a comprehensive program that is too superficial will not achieve results in any of the areas. Therefore, it is critical to base the program mix on the company's philosophy, goals, and priorities for wellness. Breadth and depth trade-offs are most easily resolved when considering the needs of both the sponsoring organization *and* the population that will be served.

Program Lines (Breadth).   The program mix is composed of different program lines that promote lifestyle change across the dimensions of wellness. Within each program line there could be a variety of options. For example, in a fitness program line, there could be aerobics classes, personal training, walking clubs, and a corporate running team. In a smoking control program line, the activities to assist employees with smoking cessation might include lifestyle courses, support groups, self-help kits, nicotine gum, acupuncture, a no-smoking policy, and financial rewards for successful quitters. Defining the breadth of a program involves selecting the areas of wellness that the program will cover. The most common areas are fitness, nutrition and weight control, stress management, smoking cessation, and preventive health screenings, although recently the concept of worksite wellness has expanded to include employee assistance, career development, and even activities like financial planning under the wellness umbrella. In fact, wellness is such a broad concept that almost any employee improvement project could be considered for the program mix.

Levels of Intervention (Depth).   There are four levels of intervention common in the health/fitness industry:

1. Communication and awareness programs
2. Screening and assessment programs
3. Education and lifestyle programs
4. Behavior change support systems

Table 2.2 presents a matrix of these four levels across five program lines (preventive health, nutrition and weight control, fitness, stress management, and smoking cessation), illustrating specific program choices. This table depicts the concepts of depth and breadth and provides a menu from which companies can select program options to configure a program mix that suits their setting.

The first level, communication and awareness programs, provides employees with information they can use to improve their health. These programs, modeled after mass-media advertising and social marketing techniques, use the "reach and repeat" concept (i.e., a brief, catchy message is intermittently displayed through a variety of stimuli) to promote specific health concepts.[9] They are intended to heighten awareness and generate further interest by providing a battery of messages around a given theme. In the mid-1970s, Stanford University showed a reduction in cardiovascular risk in cities that were

Table 2.2   Matrix of Program Lines and Levels of Intervention

Levels of intervention	Program lines				
	Preventive health	Nutrition and weight control	Fitness	Stress management	Smoking cessation
Communication and awareness programs	Newsletters Health fair Check stuffers Electronic mail Posters Fliers	Nutrition games National nutrition month events Newsletter articles Table tents	Fitness events Newsletter articles Posters Fliers	Newsletter articles Posters Fliers	Newsletter articles Posters Fliers
Screening and assessment programs	Blood pressure screening Health risk profile Cholesterol testing Health physicals Cancer screening	Nutrition assessment Computerized diet analysis Body-fat testing	Cardiovascular risk appraisal Fitness testing Body-fat testing Blood pressure screening	Biofeedback Blood pressure testing Psychoanalysis Stress questionnaires	Smoking risk assessment Carbon monoxide testing Pulmonary testing
Education and lifestyle programs	Seminars AIDS education CPR classes Physician referral system Community referral system Self-help kits	Seminars Weight loss contests Weight loss courses Cooking classes Nutrition counseling Cholesterol programs	Seminars Exercise prescription Healthy back classes Personal training Aerobics classes Walking clubs Fitness contests	Seminars Stress management workshops Time management workshops Lifestyle courses Massage therapy Psychotherapy	Seminars Support groups Behavior modification courses Hypnosis Acupuncture
Behavior change support systems	Incentive system Goal setting Resource center Buddy system	Cafeteria programs Healthy vending machines	On-site fitness center Exercise equipment Exercise trails Corporate sports teams	"Quiet" room Career development counseling Job satisfaction strategies Employee assistance programs	Nicotine gum Smoking policy No-smoking areas Smoke-free worksite

exposed to a community health education program based on mass-media intervention.[10-13] These risk reduction strategies have recently been applied to a wide range of lifestyle-related health problems in corporate and community wellness settings.[14-17]

Examples of awareness-building techniques include posters, fliers, pamphlets, newsletters, and check stuffers. Many of the communications primarily involve the distribution of printed materials, but other channels can be used (e.g., electronic mail, closed-circuit television, PA systems, telephone voice mail). These written and mass-media communication systems are frequently used to market the program, but they can also be used to introduce and reinforce positive health changes. In the design of a health communication program, it is important that the message is "framed" in a provocative manner. For example, a popular poster that was used several years ago to influence the public's perception of smoking featured an old man with missing teeth and a cigarette hanging out of his mouth, accompanied by the caption "Smoking is cool." The framing of the message is what catches people's attention and makes them think.

It is helpful to time these efforts to coincide with national themes (e.g., February is Heart Month) so employees are exposed to similar messages from television, newspaper, and radio. Many national organizations have toll-free phone numbers for consumers to call for more information—these resources can be incorporated into the program to broaden employees' access to credible information networks. When designing communication and awareness programs, the following guidelines might be useful:

- Promote a *specific theme* using simple message points.
- Use a *variety of communication systems*.
- Time the messages at *frequent intervals*.

To illustrate this concept, the following strategies could be used to promote a decrease in fat consumption during National Nutrition Month (March):

- Mailing a wallet card with paychecks at the beginning of the month that has tips for selecting low-fat foods in restaurants
- Programming the computer network to flash a nutrition message when employees log onto the system
- Circulating informational fliers on a weekly basis
- Placing table tents in the lunch room or cafeteria
- Featuring articles and sidebars that give specific guidance on cholesterol-lowering strategies in a monthly newsletter
- Publishing the American Dietetic Association's toll-free phone number to promote their consumer information center (see Appendix D)

In this example, one specific nutrition goal (e.g., to decrease fat consumption) rather than a broad message (e.g., to follow the U.S. Dietary Guidelines) has been selected. Practical information is disseminated through a variety of media. The messages have been timed to expose employees daily and weekly to reminders of how to achieve this goal. The communication program also ties into a national campaign and a professional association.

The second level of programming, screening and assessment programs, is intended to identify past, current, and potential health problems. These programs heighten awareness of health concerns because they personalize the problem to the *individuals* and then focus on their need and readiness to change.[18] For example, people might become *aware* of the dangers of high blood cholesterol through a communication program, but through a screening program they can find out if *their* cholesterol levels are too high. Most screening and assessment programs are personalized, offering participants direct contact with staff, and therefore they tend to cost more than communication-based awareness programs. General screening and assessment programs, like health risk profiles, medical histories, and physical exams, look for a range of potential problems. Other screening and assessment programs evaluate specific health risks (e.g., cholesterol, blood pressure, cancer, and cardiovascular risk screening). These programs can be offered as one-time opportunities or at periodic time intervals to provide feedback on changes that occur over time. Because assessment and screening programs only focus on identifying problems and the need to attack the problem, they fall short in providing the employee with solutions or suggestions. A screening program is most effective when used in conjunction with another level of intervention, such as a communication and awareness, or education and lifestyle, program, to strengthen the likelihood of behavior changes.

The third level of programming, education and lifestyle programs, furnishes solutions to the problems identified in the screening and assessment process. Both education and lifestyle programs give information and guidance on how to make behavior changes, but an important distinction between them is that lifestyle programs include behavior modification strategies to motivate and reinforce positive changes. Behavior modification techniques include goal setting, monitoring, cue management, problem solving, and social support.[19] This approach can be used effectively to provide individuals the specific skills necessary to initiate and maintain a behavior change, and it is very useful in weight control and smoking cessation programs.[20,21]

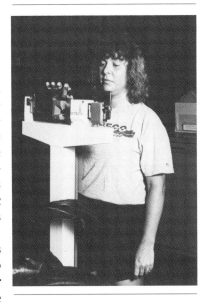

Frequently, lifestyle programs are offered as 8 to 20 week courses designed to reshape habits over a period of time, using a small-group setting (e.g., 10 to 25 participants). Educational programs, on the other hand, are usually offered as one-time seminars to either small or large groups. Other methods of promoting lifestyle change include full- or half-day workshops, individual counseling sessions, self-help kits, and a referral network to local resources (e.g., psychological counseling, weight loss programs). Many computer programs are available to analyze screening data and to design personalized health prescriptions for the employee to follow. In addition to AFB's annual directory,[3] Appendix D includes a list of other product catalogues that include vendors of education and lifestyle programs, as well as free and low-cost education materials.

The last level of programming is the development of a behavior change support system, referring to changes in the work environment that encourage employees to practice a healthy lifestyle. The other intervention techniques focus on the employees; these strategies deal with the company itself—both the corporate culture and the physical setting. This level of programming was developed from organizational development theory[22-24] and diffusion theory.[25] The goal of organizational development is to improve organizational effectiveness through interventions directed at organizational processes and worker behaviors. Diffusion theory defines processes that communicate an innovation throughout an organization. Both of these theories provide processes that help initiate and maintain a behavior-change support system. Without a behavior-change support system, only motivated employees will apply health-related information to their lifestyles. For example, it is easier to eat greasy food in the cafeteria than to walk across the street for a salad, if these are the choices available. Changes in the physical environment, such as introducing healthy menu items in an employee cafeteria or creating a smoke-free workplace, might be part of the first steps initiated in raising the corporate culture toward a higher health consciousness.

Sometimes the corporate culture grows out of the pressures in an industry. In some industries, the norm is to work 12-hour days, and a company cannot remain competitive if that culture is not adopted; it would be contradictory to introduce a stress management program when balanced lifestyles are very difficult to maintain given the long workdays. In fact, employees may even become angry when a wellness program is introduced if their work loads are so demanding that they do not have time to participate. One example of how to combat this inconsistency is to encourage employees to take an exercise break during long workdays as a stress-reduction technique, and to structure personnel policies so employees have flexibility to take their breaks when they need them (flextime). These subtle changes start to affect the cultural norms of the company.

A behavior-change support system should enhance the other interventions, adding conditions that will influence a positive health change. The following examples illustrate support systems a company could install:

- Policies that allow employees the time and freedom to participate in wellness offerings
- An incentive program for participation and goal achievement
- Recognition of positive health role models (e.g., through newsletter articles, awards program)
- Peer group activities (e.g., walking groups, a buddy system, overeating support groups, single parent groups)
- Health-oriented facilities (e.g., an on-site fitness center, healthy choices in the cafeteria, a no-smoking policy)

Small- Versus Large-Company Focus. Using the specific program ideas listed in Table 2.2 as a menu for designing a program mix, a company should reflect on its program philosophy and decide about the breadth and depth of programming. Once that focus has been determined, specific program ideas are easier to select. To illustrate how the program mix can be tailored to different settings, three hypothetical case studies are presented in Appendix E. These cases focus on how companies of various sizes have chosen their intervention strategies, and how these choices can affect the program's success.

The small company in Case 1 is developing a comprehensive program covering all the dimensions of wellness, but it is investing its more limited resources in the third and fourth levels of intervention (e.g., lifestyle courses, informal walking and jogging groups, a no-smoking policy, and a small exercise room). Although it has a broad/ shallow program mix, the support systems from level four are critical because they are fostering a health-oriented corporate culture. Key to this program's success are the commitment and active involvement of the CEO. Because she exercises at noontime, informal walking and jogging groups have formed as a lunchtime culture in the company. She is the program champion, spearheading and approving most of the wellness initiatives. An administrative assistant in the human resources department serves as the wellness coordinator and is responsible for promotions, enrollment, and vendor relations. This company has good participation rates in all the programs, especially considering the fact that they have minimal supervision from health/fitness professionals. One weakness is that the VP of human resources is not invested in the program; consequently, the program coordination is given low priority, and it takes a long time to initiate marketing activities and summarize results.

The medium-size company in Case 2 is at an early stage in program development. This company has conducted a pilot study on the training and development department and has primarily invested in level-two programming, screening and assessment programs. The program champion is the VP of training and development, who, although competent, has not been released from other duties to manage the start-up process. Consequently, the employee committee has been given responsibility for decision making and program coordination. With no consultation from technical experts, the committee decided to launch an extensive screening and assessment program, which is a labor-intensive and costly program choice. Most of the budget was spent hiring local vendors to conduct the assessments, so there was virtually no follow-up programming to support the employees in making the behavior changes suggested in the screening program. This program is caught in a common problem—lack of support from top management stifled the program's start-up, which in turn perpetuated their lack of support because the pilot study was poorly designed and the results were not very convincing.

The large, mature program presented in Case 3 has significant breadth and depth at the corporate headquarters and is focusing on level-one programming in the remote locations. The corporate headquarters' program has been based on culture-change principles that have deeply penetrated the employee population at the headquarters. The program features a 10,000-square-foot fitness center, comprehensive health education program offerings, and a health-oriented cafeteria. They have had a smoke-free workplace for 8 years. With a program of this sophistication at the corporate headquarters, the issue of fairness pushed management to consider what health promotional services could be provided at remote locations. Finally, a pilot program that consisted of a health risk profile and group health orientations was approved for 45 remote sites. Six major health awareness campaigns were launched during the pilot year, and a health newsletter was mailed to each employee's home. After a year of awareness programming, employees completed another health risk appraisal and a participation survey. The results showed that the remote-site employees were eager to learn about and initiate healthier behaviors. The company is now implementing an awareness program at all remote-site locations.

Perhaps the most important factor in designing a program mix is that it must be flexible and allowed to mature as the program moves from initiation to maintenance. Even in a large company that has extensive wellness plans there will be resource limitations, and the company may not be able to initiate all the program possibilities at the onset. Because a wellness program might be better received if it is evolutionary, rather than revolutionary, in approach, it can be beneficial to phase in a program. For example, one company may start with a fitness component and then add other elements, such as nutrition, weight control, smoking cessation, and stress management. Another company might initiate a clean-air campaign with a no-smoking policy and a series of smoking cessation programs, then later move into the other dimensions of wellness. A basic challenge for program planners, regardless of the size and scope of the program, is to maintain a program mix that consistently provides well-organized program opportunities to the employee. Uniformity in program quality, image, and philosophy will help to build a successful overall program.

Marketing Strategy

After determining what programs will be introduced to the employees and in what order they will be launched, the planning team needs to develop a marketing strategy. The marketing activities focus on two levels: the first, and most obvious, is the employees, and the second, and equally important, is top and middle management (to ensure ongoing support). Initially, the planning team will need to define its target market(s), create an image, and develop a promotional campaign.

Target Market Definition. All eligible employee groups are part of the target market. Within the total employee population, subgroups (e.g., shift workers, managers, line workers, employees in remote locations, spouses, and retirees) should be identified to focus the marketing campaign on their specific needs.[26] The results of the needs analysis will assist the planning team in learning important facts about their target markets. If possible, the employee interest questionnaire should be analyzed by employee subgroups to determine what concerns and needs are distinctive to each segment of the target market. Before conducting a broad-based promotional campaign, it is important to consider the program availability for each employee group. For example, a lunchtime lecture series will not be accessible to workers on the night shift. It would not be wise to promote the lecture series to the night workers unless the classes will also be made available at a convenient time for them. Similarly, certain communication channels work well for some segments of a work force but not for others. For example, posters and bulletin boards may not be seen by sales staff who work in home offices and seldom visit the corporate headquarters. These factors, along with any special needs of each employee group, should be considered when developing a marketing strategy.

When targeting the management, the marketing plan should incorporate the results of the management interviews. Specific steps can be planned into the program implementation and evaluation stages to keep the management involved and informed. Periodic reports and presentations, as well as special program offerings, will help to maintain and build the management's support.

Image Development. The marketing materials should reflect the program's image and project an identity that will be associated with the program for a long time. It is worthwhile to invest in designing an image that graphically portrays the philosophy of the program, yet fits the overall corporate image. Typically, programs are given a name and slogan, and a logo is designed. If an internal graphics department is available, the planning team can obtain consultation in the development of an image and color scheme for the program. Once a variety of images or logos has been developed, the employee committee can review the options and make suggestions. Top management should also have input into the final decision, because it is important that they feel some ownership in the program and can identify with the image it projects. This logo or image can be incorporated into all materials developed for the program, such as

- stationery,
- program brochures,
- fliers and posters,
- registration and testing forms, or
- incentive merchandise (e.g., T-shirts, headbands, shoelaces, gym bags).

Promotional Campaign. A series of promotional techniques will be necessary to market the program effectively. These promotions can be staggered to unfold a series of activities that build awareness, increase enrollment, and, over time, increase participation in the program. Beginning with a broad-based promotion announcing the program, such as a letter from the CEO to every employee, the marketing activities later must hone in on specific audiences and program offerings. Considering the particular needs and communication channels for reaching specific segments of the employee population, special announcements (e.g., posters, fliers) can be posted in various work areas, separate memos can be mailed, and (if the technology is available) notices can be widely disseminated through electronic mail.

Before beginning target market promotions, the individual(s) responsible for marketing should schedule appointments with managers and supervisors to address any concerns they may have about the program or specific ways they want their employees approached. For example, supervisors in production areas often are very concerned that the program might distract employees, and want program events to be scheduled at specific times, such as during breaks or lunchtime. If the management group is resistive to the program, it may be worthwhile to offer a special program, such as a management retreat, to motivate them to participate. Involvement of the CEO will be critical in shifting their attitudes.

Table 2.3 outlines various promotional techniques appropriate for both management and employees. One effective marketing strategy is to have the health promotion staff make presentations to employees, highlighting the features of the program offerings and providing information on how to enroll. These presentations should be motivating—it

Table 2.3 Promotional Techniques for Management and Employees

To management	To employees
Presentations at management meetings	Written communications
Meetings with supervisors	Letter from CEO
Management health retreat	Fliers/posters
Status reports on program development	Memos
Reports on program results	Brochures
	Check stuffers
	Special events
	Kickoff event
	Health fair
	Open house
	Keynote speaker
	Presentations to employees
	Employee referral system

is important to have an inspiring speaker who is a role model of fitness and health.

Once or twice a year, companies may want to plan special events that create excitement and high levels of participation. For example, a health fair with booths positioned in the hallways or cafeteria creates visibility for the program. These events should feature activities in which the employees can participate, such as biofeedback or risk screenings. Raffles and prizes will also draw employees to the event.

It is important to reach employees at their work stations, rather than expect them to go out of their way to hear about a program or event. In-house publications are one way to do this. Another way to do this is to have a program representative walk through work areas, introducing the program and answering questions. An employee referral program can also be effective, because it promotes word-of-mouth advertising for the program. Participating employees can receive prizes and gifts for bringing friends to the program. Finally, the employee committee members can serve as ambassadors, promoting the program within their work groups.

Staffing Model

As discussed earlier, professional staff trained in health/fitness can be a critical success factor for any health promotion program regardless of size.[5] A health/fitness professional is someone who has completed a minimum of a bachelor's degree (and desirably a master's degree) in health education, nutrition, exercise science, nursing, recreation, or allied health, and who plans a career devoted to health and fitness. These professionals help employees make safe and successful decisions about the behavior change process and provide feedback about their progress. Without input from professional staff, programs often do not proceed beyond the awareness level of programming; individual progress is slow, and the dropout rate is high, because only a small number of people are self-motivated in making health changes. For example, it has been shown that participation rate increases with the level of professional staffing, and average fitness benefits are directly proportional to participation rate.[27]

Even a small business should involve professional staff. A small business may want to contract or hire a health/fitness professional to serve as the program coordinator in charge of program design and delivery and use its existing administrative and support staff to supplement his or her efforts. At the very least, the small business may free the champion or another staff member to perform the program coordinator's responsibilities, with the employee committee serving in a support capacity (see Figure 1.1 on pp. 8-9).

A staffing model will include the staffing levels for the proposed program offerings, the type of staff required, and whether the company will hire permanent employees or contract with an external provider to staff the program.

Staffing Level. The program mix dictates the staffing needs to the greatest extent. Small companies may not have enough employees to warrant professional supervision. In this circumstance, such companies can consider cost sharing with other nearby organizations, sharing facilities in an office park, or subsidizing memberships in a YMCA or commercial fitness center. These options can help small companies provide their employees with the advantages of a fitness center, but they do not help companies address other aspects of wellness.

If a company is building a fitness center or planning a broad/deep program mix, full-time professionals and support staff will be needed to operate the program. For example, a recent survey of 68 corporate health/fitness programs (with average eligible employee populations of 800) showed that the average number of health/fitness professionals was three, with a support staff of one.[28] Most of these on-site fitness centers also provided extensive health promotion programming, including health risk appraisals, nutrition courses, weight loss courses, smoking cessation programs, and stress management workshops. If a program has a more narrow and shallow program mix, it may be possible to decrease the staffing level.

A major factor that influences the staffing level is whether the program will include a fitness facility or be more education-based. Most fitness centers are staffed with a fitness professional during all hours of operation. This staffing level is driven by facility coverage. Small facilities could have facility coverage only during peak hours of operation. A full-time employee who has floor coverage during peak hours and who offers wellness classes at other times could fill this position. A company could also hire a part-time contractor to work at the facility during peak hours. Another factor that influences the staffing level is the number of eligible employees who will participate in the program. The range for a facility-based model is generally one to five fitness professionals per 1,000 eligible employees.[27] This is a wide range because the smaller facilities require a higher ratio of staff to employees, and in a larger program the ratio tends to be lower because the facility operations gain economies of scale.

The staffing level in an education-oriented program is primarily based on the number of programs being offered because these determine the scope of work involved in coordination, promotion, and delivery. A large educational program may require a full-time health educator to coordinate the program and conduct some of the educational sessions, whereas the education-based program design lends itself to a contract or part-time staffing model, because professionals can be hired for specific programming phases. Freelance professionals may be available to deliver specialized programs, such as nutrition and stress management. Many dietitians and psychologists in private practice will come on-site to teach courses and counsel employees, and some local hospitals may also be willing to supply professional staff for program delivery.

In large programs, whether they are facility-based or program-oriented, support staff are critical to the success of the program to assist with administrative activities (e.g., development of materials, mailings, program registration, scheduling meetings, answering phone inquiries). Thus, the health/fitness professional can spend more time with the employees in program-related activities. Large programs that employ several professional staff members typically use a ratio of one support staff for every two to three health/fitness professionals.[28] In smaller programs, the support functions can be delegated to a secretary who has other duties in the company, sharing the cost and time.

Many health promotion programs administer an internship program. An internship program offers students an opportunity to earn college credit while obtaining valuable practical experience. The program can benefit by giving non-supervisory-level staff an opportunity to gain supervisory experience through managing student interns. The students provide fresh programming ideas and can help recharge old programs.

Additional details on qualifications, recruitment, and selection of professional staff is presented in more detail in Phase III, Implementation (p. 66).

Hiring or Contracting for Staff. An important decision in developing a staffing model is whether the company should hire its own staff or contract with an outside provider. Some companies want total control over their programs and recruit qualified program directors to run them. Other organizations would rather leave this responsibility to a competent provider with a proven track record in serving other companies.

Many companies choose the route of hiring full-time program staff and use a health/fitness consultant to assist with hiring the program director, who, in turn, assumes the responsibility for creating the program technology, managing operations, and hiring other staff. Companies that choose this option feel strongly that full-time employees have more ownership in the program, and that the health/fitness professional will become more integrated into the corporate culture. They are also willing to commit the management time to learn about the health/fitness industry and to oversee the director's work.

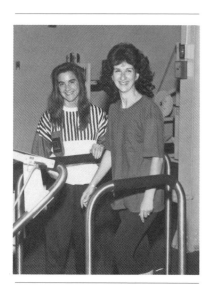

A full-time program director in a medium to large program that will employ two or three full-time professionals and one support staff should have a minimum of a master's degree and 5 to 10 years of experience in the field. Smaller programs that will employ only one employee should hire a professional who has at least 5 years of experience and at least a bachelor's degree. The health/fitness professional's degree (whether it is a graduate or an undergraduate degree) should be in a field that directly relates to the program's focus (e.g., nutrition, health education, fitness). If extensive program development is involved, it is important to hire a more experienced professional, even if the program is small. Also, it is necessary to allocate sufficient time during start-up for the program director to design the program, purchase materials, and develop the protocols.

In other cases, a contract management company can provide the staff resources and program technology (e.g., protocols, program materials, operations manuals). Most providers will customize these materials, adapting procedures to fit the setting and employee population and incorporating the image of the program into the program materials. The provider should supply the company with continually updated technology. This saves development time and frees the professional staff for program marketing and delivery. Because the provider assumes responsibility for recruiting, training, supervising, disciplining, and ongoing quality control of staff, the sponsoring company is free from managing a department with goals and objectives different from their major line of business. This approach works well in small- to medium-size programs that do not require full-time staff, because a contract management company can place full-time staff in more than one setting, spreading their time across a couple of contracts. It can be very difficult to find experienced staff on a part-time basis, and sometimes the only way to staff a program on a part-time basis is with entry-level professionals or students. In these cases, a management company can serve a useful function in hiring, training, and supervising these less experienced professionals.

A middle course used by some companies is a hybrid. They hire a program director, but use a provider organization as a resource for program technology, supplemental professional staff, and research. Once again, this decision is influenced by the philosophy and scope of the program, the program mix, the number of eligible employees, the availability of local resources, and what programs and facilities are being planned.

Job Classifications.   The number of staff, job titles, and responsibilities vary according to the size and complexity of the program. In small programs, the program director is responsible for everything from program delivery to collating packets for a mailing. In large programs, support staff are available to assist with administrative duties. Job classifications also vary depending on the program focus (i.e., whether it is a facility- or an education-based program). Appendix F presents job descriptions, qualifications, and salary ranges for both professional and support staff. These job descriptions relate more to a large, facility-based program, but they can be modified to fit other program models. For example, in a small program the duties that are spread across several of the job descriptions need to be consolidated into the program director's job description.

The program director is ultimately responsible for all aspects of program development, marketing, implementation, and evaluation. If a company decides to hire a full-time program director, this individual should be hired during Phase II, Conceptual Definition, so that he or she has input into the overall program design. In large programs, an assistant director may be hired, and some of the director's duties will be delegated to the assistant. The assistant might have an advanced degree, but does not need to have the same level of experience as a program director.

Fitness professionals, sometimes referred to as fitness counselors, exercise leaders, or exercise specialists, usually serve as the on-site staff in a fitness center. They also may provide freelance services to conduct aerobics classes, fitness assessments, or seminars on exercise topics. In small programs these individuals could be hired as part-time employees to manage the facility during peak hours. They would also orient employees to the facility and equipment. These staff members should have a minimum of a bachelor's degree and ideally some level of certification. (Refer to pages 66 and 67 for a more in-depth discussion on certifications.)

A health educator would be responsible for planning, coordinating, and delivering health education programs. In a broad-based program that offers many level-one, -two, and -three programs but does not have a fitness center, a health educator may be a more appropriate program director than a fitness professional. In this case the management duties delineated in the program director's job description can be incorporated into the health educator's responsibilities. Similarly, a nurse might be more appropriate for a program core based on screening and assessment. In most facility-based programs, health educators supplement the fitness staff on a part-time basis. Some health/fitness professionals have cross-trained, acquiring combination degrees (e.g., in exercise physiology and nutrition, health and physical education).

These individuals are ideal for programs of all sizes, because the professional can perform in multiple roles.

Support staff could be an administrative assistant or, in a facility-based program, a facility attendant. An administrative assistant not only provides clerical support, but also serves as a receptionist, maintains inventory, orders supplies, processes program fees, and maintains program records. In large fitness centers, it is helpful to hire facility attendants to supplement professional staff during peak hours of use (e.g., noon-hour rushes). These individuals perform routine housekeeping duties and monitor the front desk. An undergraduate student interested in working in a worksite health/fitness environment could fill this position.

Facility Plan for Companies Developing Fitness Centers

A major focus in developing a program proposal is the development of a facility plan that fits the overall program design. If a fitness center is being proposed, the facility plan will be more detailed because it will involve significantly greater capital and operating costs. Each program component should be evaluated for the estimated number of employees it will serve, functional space requirements, equipment needs, and overall priority within the program. By prioritizing program components, a wish list of facility and equipment needs and the associated initial cost and maintenance cost will be formed. At this stage, it is also important to look for activities that compliment each other. This provides the potential for multiple-use design and could reduce the total square-footage requirements. This section discusses space requirements and layout and design considerations.

Space Requirements.   The first issue is to determine the total square footage of the facility. Facility size should be geared to the projected peak-hour usage, which will require a detailed analysis of the eligibility criteria, employee demographics, work schedules, and transportation patterns. Square-footage requirements per eligible employee will vary depending on the size of the eligible group as well as the types of facilities to be installed. For example, swimming pools and racquetball courts require more square feet per member than do aerobics studios.

One company, Fitness Systems, obtained data from 40 existing facilities to examine current square footage norms and trends. Their research showed that the mean square footage for a core fitness facility ranges from 2.2 to 6.7 square feet per employee, depending on the size of the employee population. In general, as the number of employees increases, the square footage per employee decreases.

Once the total size has been estimated and potential locations identified, it is necessary to determine the size of functional areas within the facility. The functional space requirements depend on the program mix and projected participation rates for each program component, as well as on the overall facility size. Table 2.4 presents averages used by one company for predicting the types of exercise areas included in facilities of various sizes. Note that the larger facilities (over 15,000 square feet) can include less space-efficient facilities, such as basketball and volleyball courts, racquet courts, and swimming pools. These facilities add variety to programming options and entice more employ-

Table 2.4   Percentage of Corporate Fitness Centers That Include Various
Exercise Areas

Exercise areas	Facility size (square feet)		
	< 7,500	7,500-15,000	> 15,000
Equipment room	100%	100%	100%
Movement class area	76%	92%	100%
Indoor track	8%	62%	73%
Volleyball or basketball	4%	38%	73%
Racquetball or handball	0%	23%	55%
Indoor pool	0%	0%	55%

Note. Survey of 49 companies compiled by Fitness Systems, 1985.

ees to participate. Obviously, including these more expensive facilities
is dependent on the program philosophy and budget.

Table 2.5 presents the typical space allocations required for a basic
employee fitness center. The main exercise area is the largest space
within the fitness center. It should furnish participants with many
different exercise options, including equipment for muscular strength
and endurance work and cardiovascular conditioning. To predict a
facility's capacity, estimate the number of participants who will use
the facility during peak usage and add a 10% cushion.[29] The main
exercise area should allow 40 to 50 square feet per peak-hour
participant.

The aerobics studio should have 36 to 40 square feet per participant
in an exercise class at any given time.[29] The flooring should be a dura-
ble and easily cleaned surface that is designed to minimize the potential for injury. Many of these rooms use mirrors that make the room
seem bigger and brighter and help the instructor observe the participants' body alignment. Some studios include special lighting that enhances the participants' comfort and enjoyment. The air circulation in
these rooms must allow for air exchange with outside air. A proper
sound system should be installed to project music and the instructor's
verbal cues.

Table 2.5   Recommended Space Allocation in a Fitness Facility

Area	Range of space allocation
Main exercise area	40-50%
Locker rooms	20-35%
Circulation areas	5-15%
Aerobics studio	5-10%
Administration	5-10%
Storage	2-4%

Note. From *Implementing Health/Fitness Programs* (p. 268) by R.W. Patton, J.M.
Corry, L.R. Gettman, and J.S. Graf, 1986, Champaign, IL: Human Kinetics.
Adapted by permission.

Second to the main exercise area, locker rooms require the largest amount of space. The locker rooms are an important auxiliary area that affects the atmosphere and image of the entire facility. Locker rooms have areas for dressing, showering, and drying. They should be designed so that the traffic flow allows individuals to move in and out of these various areas comfortably. The locker areas should be convenient to the exercise area, ideally placed in a logical traffic flow between the front desk and exercise areas. The amenities provided in the locker rooms will vary according to the philosophy of the program. Some amenities that are seen in worksite fitness centers are these:

- Towels
- Soap and shampoo
- Hair dryers and curling irons
- A full range of toiletries (e.g., shaving cream, razors, hair spray, foot powder)
- Exercise clothing
- Laundry service for exercise clothing
- Storage bins or kit lockers

There is a need for administrative areas regardless of the facility size. Table 2.6 lists the administrative areas that require space (the front desk, teaching area, staff office, and storage areas). In addition, circulation space is required to allow for traffic flow from area to area. Circulation areas usually comprise 5% to 15% of the total facility space. These areas should be well lit and decorated with plants and pictures to enhance the appearance of the facility, create a pleasant atmo-

Table 2.6   Administrative Areas Requiring Space

Designated area	Activity
Front desk	Check-in Locker assignment Equipment checkout Facility security
Teaching and testing area	Health and fitness program orientation Exercise prescription and individual counseling Blood analysis Fitness assessments Health education classes Self-help library Interactive computers
Staff office	Staff work area Record/filing area Copy machine Computer(s) and printer(s) Conference area
Storage areas	Program material Equipment

sphere, and soften the hard lines created by the exercise equipment. Wall space can be used for bulletin boards and instructional posters to provide information and enhance the educational experience.

Layout and Design Considerations. Once the total and functional space requirements are determined, a space planner can be hired to develop architectural drawings. Oftentimes, a technical expert in facility design is contracted to work with the architect to develop a space plan for the facility. The technical expert advises the space planners on considerations unique to a fitness center design, such as these:

- Air exchange ratios
- Traffic flow patterns
- Space requirements
- Wet and dry areas in the locker rooms
- Wall coverings and carpeting
- Floorings
- Front desk placement
- Placement of computer terminals
- Placement of equipment

Most corporate fitness centers operate with a lean staff, so it is important to consider the labor intensity of the facility design. The program mix and staffing model are therefore very important factors to consider in developing a facility plan. For example, a facility designed for operation by one staff member should have the front desk placed with an unobstructed view of the main exercise area. The administrative areas should be adjacent, so the staff member can efficiently perform multiple duties during low-usage times (e.g., midmorning and midafternoon). He or she should be able to monitor the front desk, observe the exercise area, answer the phone, put up a bulletin board, and handle routine paperwork. Likewise, the facility plan should accommodate special programming needs, such as counseling cubicles, classrooms, testing areas, and so on.

Despite planning and consultation, every facility reveals some mistakes upon opening that will require creative problem solving during operations. Knowing of potential problems during the planning stage can prevent operational inefficiencies later. Some common facility and equipment mistakes are these:

- Inefficient traffic flow
- Inappropriate placement of service and activity areas
- No facilities for people with disabilities
- Inadequate office space
- Inadequate and inconvenient storage space
- Failure to plan for future expansion
- Inadequate amount, or wrong type, of exercise equipment
- Equipment requiring high daily maintenance
- Equipment requiring high staff supervision
- Inadequate open space for stretching and traffic flow

Facility Plan for Companies Developing Educational Programs

Installing educational programs at the worksite is much easier than installing a fitness center, because office space (e.g., conference rooms, private office space, storage closets) is more easily adapted to the needs of educational programs. However, this does not mean that an educational program does not require dedicated space and an investment in equipment. The facility needs for an educational program depend on the type of program (e.g., nutrition, stress management, smoking cessation) as well as the delivery method. If the program design involves primarily communication systems and self-help materials (e.g., books, videotapes) rather than courses and seminars, an extensive resource center and computer network might be the most useful investment in space and equipment. Table 2.7 presents a list of the facilities that might be required for delivering educational programs.

Conference rooms will be one of the most important facility needs for an educational program. There should be at least one room available to hold 10 to 20 people for small group sessions and committee meetings. In some programs, this will meet most of the programming needs, but to deliver large educational sessions (e.g., for 20 to 100

Table 2.7 Facility Needs for Educational Programs

Facilities	Square footage	Considerations
Teaching rooms		
Conferences	600	Capacity of 10-20 seats
Large lecture halls	1,250	Capacity of 25-100 seats
Counseling areas		Privacy is important for participants to feel comfortable
Private office	120-150	
Counseling cubicles	75	Space could be shared
Self-help resource area	200-400	Bulletin board Books Videos Tapes
Special programming space	600-1,250	Depending on program focus, space could be dedicated for special programming
Quiet room		
Demonstration kitchen		
Health assessment area	200-400	Blood drawing Height/weight Fitness tests
Storage room	200-400	Printed materials Teaching aids Portable equipment
Administrative/office space	120/full-time staff member	Professional staff Support staff

Note. Data represents space utilization requirements compiled from Tenneco Training Center and Health Promotion Program.

participants) and gain economies in professional time, an auditorium-style room will be necessary. Many modern office buildings offer auditoriums as an amenity; corporations located in these settings can rent the auditorium for these programs.

Teaching rooms should be well ventilated, with adjustable temperature controls. A room that is too hot and stuffy, or too cold and drafty, can dramatically affect the participants' comfort and attention span. The seats should be comfortable with good padding and back support, especially if participants will be involved in full- or half-day workshops. Some workshops will require tables for participants. The lighting system should provide the flexibility to darken the room sufficiently for slide projection, but also supply indirect lighting so participants can write notes. A teaching room should have good acoustics and no visual obstacles (e.g., pillars that block the view from behind), so all the participants can see and hear the entire session.

The professional staff should have access to a private office for counseling program participants, even if the professionals are not full-time employees. To save space, specially designed counseling cubicles could be installed. Dedicated space for the health promotion program enhances the creative programming possibilities. For example, a demonstration kitchen, a quiet room for practicing relaxation techniques, or a self-help library could be installed. Every program needs adequate storage space for program materials, fliers, brochures, supplies, and equipment. The storage area should have sufficient shelf space to organize and label materials. To prevent damage and pilfering of supplies, it is a good idea to store materials in a locked area.

Equipment Needs

Programs of all sizes have some equipment needs, from scales to a full line of exercise equipment for a fitness center. With the proliferation of exercise equipment on the market, selecting exercise equipment can be a challenge to program planners. Educational programs require audiovisual (AV) equipment and other teaching aids. Administration of most programs is facilitated by access to computers and other office equipment. The equipment needs for fitness centers and educational programs are addressed in this section.

Fitness Equipment. Table 2.8 presents a checklist of considerations for selecting exercise equipment. Durability, safety, and simplicity are important considerations that affect the performance of the equipment in the facility. The manufacturer's support services, documentation, and reputation are very important to the maintenance program and consequently have a lasting impact on the facility operations. A cheap piece of equipment that constantly breaks and is costly to repair will be more expensive in the long run.

There are specific considerations for cardiovascular and strength-training equipment. Cardiovascular or aerobic conditioning is usually the core fitness objective of worksite health/fitness programs. This requires a variety of aerobic conditioning equipment for the main exercise area—Table 2.9 outlines some of these options.

Even though cardiovascular conditioning may be the core fitness activity, strength training has gained popularity over the last 5 to 10 years, and every facility devotes part of the main exercise area to strength-training equipment. Strength-training equipment can be

Table 2.8   Checklist for Selecting Exercise Equipment

Simplicity

Is it self-explanatory to use?
Does it require minimal adjustments?

Durability

Steel or aluminum?
Will it stand securely on its base?
Will it hold up with heavy use and user abuse?
What is the recommended life?

Safety

What safety documentation does the manufacturer provide?
Are moving parts enclosed to protect users?
Was the unit lab tested to determine the safety limits?
What safety features does it have?

Support

Does the manufacturer have a good reputation?
Are there regional vendors who maintain inventories for parts and/or
    service people?
Is there a prompt delivery schedule?
Is there a warranty? What does the warranty cover? How long is the
    warranty?

Appealing

Will it look good in the facility?
Is it comfortable and does it feel right?
Do members of other facilities use the equipment regularly?

Documentation

Are the design and performance standards explained in an owner's
    manual?
What research has supported the design and testing?

Company Reputation

Are other fitness centers happy with the quality, service, and durability of
    the equipment?

divided into two basic categories: machines and free weights. Both are very popular for a variety of reasons, yet they pose significantly different safety challenges. Proper free-weight use requires spotters (who might be professional personal trainers or exercise buddies) to ensure that participants are lifting properly and are not injured. Extensive use of free-weight equipment may require a higher staffing level to supervise training programs. With weight machines, major concerns are body alignment and proper lifting technique. Most worksite fitness centers purchase strength-training machines that are ergonomically designed to place the body in appropriate postures and provide support, so that once participants are instructed in proper lifting techniques they can train safely by themselves. The machines, therefore, minimize the risk of injury and would be the preferred equipment in a facility. Machines are more expensive than free weights. Most manufacturers require a minimum of 45 to 65 square feet for each piece of equipment, so machines may not be feasible in small facilities. The machines should be arranged to promote a logical progression from one piece of equipment to another during training. The staff should also be able to observe the weight machines from administrative areas of the facility. The ratio of free weights to strength-training machines

Table 2.9   Aerobic Exercise Equipment Options

Equipment	Comments
Stationary cycles	Easy to adjust resistance Comfortable seat Easy to adjust height and position Provide a variety of models for interest and wide appeal
Treadmills	Good substitute for an indoor track Motorized treadmills vary speed, elevation Adjustments can be manual or computerized
Rowing ergometers	Offers aerobic and strength training Odometers monitor each stroke Sliding seat should be comfortable Adjustable foot should accommodate a variety of sizes
Stair climbers	Mimics stair climbing Strengthens lower body muscles Can be manual or computerized

will depend on the population who will be using the facility, as well as the program mix and staffing model.

Educational Programs.   Table 2.10 presents a comprehensive list of the types of equipment that might be used in educational programs. These needs include audiovisual equipment, computer hardware and software, health assessment equipment, and portable cooking equipment. Some of this equipment may already be available in an office environment, so it can be shared with other departments to avoid making a capital investment.

AV equipment is the most necessary type of equipment for educational programs. A 35-mm slide projector and an overhead projector are the most commonly used pieces of AV equipment for teaching health promotion. If the company already owns these items, it may not be necessary to purchase additional equipment. However, in large programs that are running more than one class at a time, or in corporate settings where the AV equipment is extensively used for other meetings, the program may need its own equipment. Also, this equipment could be rented on an as-needed basis. Video cassette recorders (VCRs) are becoming a more popular method for teaching health promotion, especially because there are now more high-quality health/ fitness tapes available. Videotapes can be a useful supplement to an educational seminar for demonstrating an activity that requires special facilities that are not available (e.g., cooking methods, shopping techniques). A video library could be useful in a self-help resource center. If videotapes will be used in a large lecture setting, it might be necessary to use a video projector that enlarges the image on a screen so everyone can see the picture.

Computer systems have many applications for educational programs. A word processing system is an important tool for every work-

Table 2.10   Equipment Needs for Educational Programs

Type of equipment	Possible items
Audiovisual equipment	Slide projector Overhead projector VCR player/recorder Video projector Audiocassette player Camcorder
Computer systems	Laser printer Word processing software Graphics software Desktop publishing software Financial software Data-base software Fitness software Health risk profile software Nutrient analysis software Electronic mail
Health assessment equipment	Scale Blood pressure cuff and sphygmomanometer Body-fat assessment equipment Biofeedback equipment Carbon monoxide analyzer Fitness testing equipment[a] Finger-prick blood draw system
Cooking demonstration equipment	Microwave Electric wok Mobile food carts Plastic food models
Office equipment	Copier Fax machine Telephone Answering machine

[a]Flexibility, strength, and cardiovascular fitness assessment.

site and should be made available to the health promotion staff. Graphics and desktop publishing software can be used for preparing educational materials, fliers, brochures, and audiovisual aids. By designing printed materials in-house, the program can have customized pieces that are graphically appealing. Laser printers produce copy that is camera-ready for many publications (e.g., newsletters, fliers, handouts, and manuals), which can save in printing and graphics costs. Financial and data-base software are useful for record keeping and program evaluation. There are many health/fitness software programs that might be useful in conducting health assessments (Appendix D provides a list of product catalogues). If the company has an electronic mail system, it can be used to disseminate program information (e.g., promotions, schedule and room changes) and also to broadcast health messages (e.g., a health "tip of the day," nutrition trivia).

Health assessment equipment is important for collecting objective data to provide feedback to the participants and measure program results. A scale and blood pressure system are useful in screening and assessment programs. Equipment to measure body composition is useful in screening, weight loss, and fitness programs. Biofeedback equipment could be used in a stress management program to evaluate the physiological responses to stress and relaxation. Similarly, a carbon monoxide analyzer can be used in a smoking cessation program to document quit rates.

Nutrition education programs are more effective if participants can watch food preparation techniques and taste food made with modified recipes. A program that focuses extensively on nutrition education might even install a food demonstration kitchen where participants not only observe, but also prepare healthy foods. Without kitchen facilities, it is still possible to incorporate cooking experiences into a nutrition education program. A microwave and electric wok are relatively easy to transport and can be set up in a conference room for a food demonstration. In many settings, the employee dining area has a refrigerator, sink, and microwave; a cooking class could be held during midmorning or midafternoon hours when this room is not being heavily used.

Financial Plan

Every program proposal requires a detailed budget of anticipated start-up and operating costs, and it will have two basic components: the capital and the operating budgets. A budget is both a planning and a controlling tool.[30] As a plan, it provides a specific statement of anticipated financial outlays and covers a specific time period. Because the budget is frequently a major determinant of the program scope, once the costs associated with a specific program design have been identified, it may be necessary to revise some of the elements of the choices, such as scope of program offerings, staffing level, facility design, and equipment selections. When a budget is properly administered, it is a tool for control and accountability, because it gives the line manager the financial guidelines to accomplish the program's goals and objectives. The capital budget might be a major concern during start-up if facility construction or renovation are being planned. On an ongoing basis, the operating budget will be the most involved financial planning tool. During start-up, both of these budgets will shape the program design and form the basis for future financial plans.

Capital Budget. The capital budget includes all items of a permanent or semipermanent nature, such as land, buildings, and equipment. Equipment in the capital budget usually refers to fixed equipment that has a value over $500 and a life expectancy of more than 2 years. When purchasing major equipment items, it is important to calculate depreciation cost by amortizing the acquisition cost over the life of the equipment. The depreciation cost might become a line item in the operating budget. An expensive piece of equipment may be more cost-effective in the long run because it might have a longer life expectancy, slower rate of depreciation, and lower maintenance costs.

Table 2.11 presents the cost ranges associated with the capital investment in a fitness center. The totals of the costs itemized on Table 2.11 vary from $70 to $170 per square foot, depending on the sophistication of the overall design, equipment choices, and finishings. Most corporate fitness centers do not require extensive engineering fees, because they are usually incorporated into an existing building. If the fitness center is planned as a stand-alone facility, the architectural and engineering fees would be significantly higher. The construction/build-out fees are influenced by factors such as retrofitting of plumbing and electricals, and changes in the HVAC system, as well as geographical location. The equipment budget will be largely influenced by the type of equipment selected. For example, high-tech, computerized equipment costs substantially more than manual equipment.

If the program mix does not involve a fitness center, the capital budget would include other necessary equipment, such as computers, audiovisual equipment, scales, and testing equipment. In most cases, it will not be necessary to incur construction costs, unless there are renovation plans to install special teaching facilities (e.g., a demonstration kitchen). Table 2.12 presents budget guidelines for an educational program. A few items that have a significant financial impact on the start-up budget for an educational program (e.g., training and development, printing and graphics, program materials) have been identified, even though they are not typically considered capital purchases. Because an educational program is much less tangible than a fitness center, a common mistake in planning these programs is to overlook the magnitude of the initial investment. The result is that the program struggles with insufficient resources and never becomes a major intervention strategy.

Many of the items listed in Table 2.12 have a wide range of costs. The costs will vary depending on the overall sophistication of equipment chosen (e.g., a top-of-the-line Macintosh computer system with a laser printer and extensive graphics software versus an IBM clone with word processing software and a dot matrix printer). Training and

Table 2.11 Capital Budget Guidelines for Fitness Facilities of Various Sizes

Type of cost	Facility size (in square feet)			
	1,500	4,000	8,000	15,000
Feasibility study/needs analysis	$1,500-2,000	$1,500-2,000	$2,500-5,000	$2,500-7,500
Architect/space planner	$3,500-5,000	$3,500-5,000	$3,500-5,000	$5,000-7,500
Construction/buildout	$50-100/sq ft	$50-100/sq ft	$50-100/sq ft	$50-100/sq ft
Construction contingency[a]	10%	10%	10%	10%
Equipment and furnishings	$30,000-60,000	$60,000-100,000	$80,000-125,000	$100,000-200,000
Legal fees	$1,500-3,000	$1,500-3,000	$1,500-3,000	$1,500-3,000
Health/fitness consultant	$2,500-5,000	$5,000-10,000	$10,000-25,000	$15,000-30,000
Licenses and permits	$1,000	$1,000	$1,000	$1,000

Note. Cost estimates based on 1990 figures.

[a]10% of the construction cost should be budgeted to allow for cost overruns (e.g., overtime, weather-related delays, and other unexpected causes).

Table 2.12 Budget Guidelines for Educational Programs of Various Sizes

Type of cost	Number of employees		
	≤ 500	1,000	≥ 2,000
Feasibility study/needs analysis	$1,000-1,500	$1,000-2,000	$1,500-2,500
Equipment			
Audiovisual[a]	$500-3000	$500-3000	$500-3000
Computer	$3,000[b]-10,000[c]	$3,000[b]-10,000[c]	$3,000[b]-10,000[c]
Health assessment	$1,000[d]-10,000[e]	$1,000[d]-10,000[e]	$1,000[d]-10,000[e]
Cooking[f]	$1,000	$1,000	$1,000
Training and development[g]	$1,000-2,500	$2,500-5,000	$5,000-15,000
Printing/graphics[h]	$500-2,500	$1,500-10,000	$10,000-30,000
Program materials[i]	$1,500-3,000	$3,000-8,000	$8,000-25,000
Self-help resources[j]	$500-1,500	$1,000-2,500	$2,000-4,000

Note. Cost estimates based on 1990 figures.

[a]Depends on the type and number of equipment purchased. [b]IBM clone with dot matrix printer and minimal software purchases. [c]Macintosh computer with laser printer and extensive software investment. [d]A few pieces of low-cost manual equipment. [e]Full line of high-tech equipment (e.g., bioimpedance analyzer, biofeedback equipment, carbon monoxide analyzer, computerized testing equipment). [f]Portable cookware and food models. [g]Training and development could range from sending professional staff to industry conferences to an extensive lay-leader training program (travel costs not included). [h]Printing and graphics covers design of image and printing of letterhead, forms, etc. If all program materials are custom printed using multiple-color printing, these costs will be higher. [i]Cost of program materials will depend on whether the company purchases or internally develops programs, and the sophistication of materials purchased. If internally developed, labor and printing/design expenses will be higher and program materials will be lower. [j]Investment in self-help resources will depend on the type of materials (e.g., videotape vs. paperback books) and how many are purchased.

development costs could be very high if a lay-leader training program was conducted to train 50+ employees from various remote locations. Printing and graphics costs are high if most program materials are developed in-house, and the budget for program materials is higher if most materials are purchased from outside vendors. Oftentimes, the cost of printing versus program materials line-items are a trade-off that reflects the program philosophy. The quality of printing (e.g., two-color, paper choice) and amount of graphic design also affect the printing budget.

Operating Budget. The operating budget consists of expenses associated with the day-to-day operation of the program or facility and anticipated revenue, calculated on a monthly basis. Because the budget provides a basis for comparing planned to actual performance, financial control is enhanced with a well-prepared operating budget. Corporate facilities may or may not have any revenue, depending on whether or not the company plans to charge the employees a fee for participating. Typically, a modest fee (e.g., $5 to $25 per month) is charged to instill a sense of commitment to the program without raising an economic barrier to participation. Some companies refund a portion of the fee after goal achievement or regular participation has been demonstrated. In a recent survey of major corporate fitness programs, about half of the programs charged the participants on some recurring basis (e.g., weekly, monthly, or annually).[28] These fees usually do not cover the entire cost of the program or facility, so the operating budget may not have a positive cash flow.

The largest operating expense will be the wages and salaries for the full- and part-time staff. Salary adjustments for merit increases and promotions, and any costs associated with temporary coverage for vacations and sick leave, should be anticipated. Benefit costs will range from 10% to 20% of total salaries for part-time employees and 20% to 35% for full-time employees, and should be factored into the total personnel costs.

A major line item in commercial facilities is rent, utilities, and occupancy costs. In most corporate fitness centers, rent and utilities are absorbed in some other area of the corporate budget, although some companies will consider rent a lost revenue opportunity and want to track this cost even though the program or facility might not be expected to actually pay for rent. Every company has standardized line items for budgeting, and the health promotion program or facility budget should conform to these categories.

Regardless of the program scope and the company's level of funding for program operations, the operating costs should be tracked and measured against the outcomes. Table 1.4 (p. 10) presents the per-capita costs for various types of wellness programs. Without appropriate budgeting and tracking, it will be impossible to conduct cost-effectiveness analysis.

Summary

In Phase II, Conceptual Definition, a program philosophy and model are created that address the corporate needs that were identified in Phase I. The activities associated with Conceptual Definition are referred to as (a) philosophy and scope, and (b) program design. Developing a mission statement and a set of goals and objectives that are realistic and measurable helps to clarify a direction for the program. At some point in the early stages of Phase II, it will be necessary to narrow the scope of the program by establishing priorities in program offerings. When defining the philosophy and scope of the program, it is helpful to develop an evaluation plan that addresses how the goals will be measured. The program design stage is the most detailed aspect of Conceptual Definition. The core and supplemental program offerings need to be selected from a menu of options. These choices usually involve making a trade-off in program breadth or depth, which should reflect the program's philosophy and goals. The program design also needs to address the marketing strategy, staffing model, facility plan, and equipment needs. All of these design decisions are interrelated; for example, the specific program choices affect the staffing needs, which in turn influence how the program is marketed. Because staffing, marketing, facility, and equipment selections all have financial implications, a detailed budget will need to be prepared. It may be necessary to revise the program design after the total cost has been estimated, to match the level of funding that has been approved for the program. Creating a program design is like assembling the pieces of a jigsaw puzzle; the design team must work on the details of fitting small pieces together, but not lose sight of the big picture.

PHASE III

Implementation

Implementation
• Program activation • Marketing and promotions • Staff selection • Operations and   administration

The implementation phase of a worksite health promotion program is probably the most critical period in program operations. During this time all the plans are activated and the philosophy of the program is initially presented to employees. This opportunity only occurs once in the life of a program, yet it can set the tone for future success or failure. Crucial to this stage is the ability to create a beginning that is exciting and challenging to employees. It should enlist support from all employees, starting a momentum that begins to impact the corporate culture.

Employee health and fitness programs are successful when high participation rates are maintained, and the health behavior change process affects large groups of employees. The positive momentum generated from these two factors builds a health consciousness within the corporate culture. A successful program initiation helps to achieve the high level of employee participation that becomes a base for future growth and development.

During Phase III, Implementation, the various plans that were developed in Phases I and II are implemented. A team of health/fitness professionals is organized according to the proposed staffing model. The program mix designed in Phase II is activated. The marketing strategy is initialized, and the various promotional and incentive techniques are launched. Facility and equipment issues are finalized, and administrative systems are established to ensure a smooth, efficient, and high-quality operation.

Program Activation

To meet the challenge of achieving high participation and adherence rates, an assortment of activities should be scheduled throughout the year. Developing a calendar of events is an essential step in program implementation because it provides a framework for planning and managing resources. During the program activation stage, enrollment and screening procedures might need to be established. Strategies to motivate employees and promote retention should be incorporated into the implementation process. All these aspects of successful program delivery will ensure that the program makes a positive first impression.

Program Calendar

Using the program mix as a basis for planning, a calendar can be used to outline plans for the various programs and special events. A program calendar is a planning tool that helps staff manage the timing issues associated with program development, implementation, and evaluation. It also serves as a useful program guide for the employees.

The first consideration in developing a program calendar is to offer an assortment of activities targeted to employees' varying needs and interests. For example, Table 3.1 presents a company-wide weight control strategy that incorporates program offerings at all levels of intervention and provides options to suit a variety of employees' preferences. The program activities can be staggered throughout the year to create continuous programming within a given program line. The schedule of events will be a major determinant of success, so it is important to think through several timing issues.

Program Length. It takes a certain length of time to see participants make positive behavior changes, but the longer the program, the harder it is to maintain program adherence. Programmers may need to experiment to find the appropriate length of time for each program offering. Many weight loss and exercise programs are 8 to 12 weeks long. In this time period, participants experience both physiological

Table 3.1 Weight Control Program Strategy

Program title	Program description	Target population
Nutrition Assessment	Ideal weight, calorie needs, nutrition adequacy.	Participants desiring an evaluation of their diet and caloric needs.
Individual Counseling	One-on-one counseling for the employee by one or more health professionals (medical, nutrition, psychology, fitness). The staff group studies each employee and tailors the program.	Employees who have more than 50 pounds to lose and need individualized guidance.
Weight Loss Courses	Group support and information program that offers continual noncompetitive support.	Employees who want or need continual support and a structured program.
Weight Loss Competition	Team competition involving education and practice of good nutrition, exercise, and behavior modification strategies. Four- to 8-week program.	Employees wanting to lose weight or maintain their body weight, and who are motivated by competition.
Staff Coaching and Counseling	One-on-one coaching and counseling for individual employees.	Employees interested in one-on-one coaching and counseling.
Brown Bag Lunch Program	Noontime lecture series offering nutrition and weight control information.	Employees who want to make a minimal time investment.
Self-Help Information	Bulletin boards, pickup materials.	Employees who are not interested in group participation and are self-motivated.

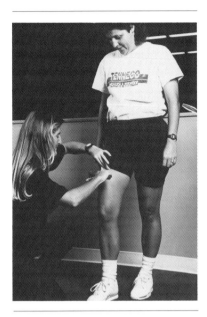

and psychological improvements, and they begin to understand the process of behavior change. Although programs that are much longer than this tend to have higher attrition rates, sessions that involve intense coaching and counseling can be longer, because of the one-on-one contact. Activities staggered throughout the year can vary in length, appealing to different populations and maintaining an awareness that supports a specific behavior change. For example, the weight control strategies presented in Table 3.1 could be scheduled at different times throughout the year.

Seasonal Cycles.   Within the health/fitness industry, there are annual programming cycles. In January, people nationwide think about New Year's resolutions. This is an opportune time to kick off new programs, particularly nutrition, fitness, and weight loss. Because many people have previously resolved to lose weight, become fit, and live healthier, but have been unsuccessful in maintaining their commitments, this presents an opportunity to teach goal-setting strategies.

The months before the swimsuit season are another time when people seek fitness and nutrition programs. Employees are eager to burn off fat and tone muscles that will be revealed in warm-weather clothing. The summer months offer many outdoor challenges, but employees find that family activities compete for their time. Worksites usually empty early during the summer months and operate at skeleton staffing levels while employees take vacations. Program participation will drop as a function of summertime demands. This is a good time for the staff to develop future program plans, collect and analyze evaluation data, and catch up on paperwork.

September starts the school cycle, and employees become more focused on work. The Labor Day weekend marks the end of summer, and people naturally fall into a new pattern in their lifestyles. This is an excellent time to introduce new habits into employees' routines. Wellness programs launched during September and October will generate an active participation and provide momentum for the end of the year. November and December are filled with holiday spirit and too much food. It is a time when participants need help and suggestions to curb the holiday bulge. A holiday weight challenge may help to combat the tendency to indulge and overdo it during the holiday months. Figure 3.1 presents a sample program calendar.

Special Events.   Throughout the calendar year, two or three special events should be planned to contribute fun, variety, educational experiences, public relations opportunities, and motivational incentives. These events can range from corporate sports teams to health fairs. Scheduling these events to coincide with national or state activities, such as the Great American Smoke Out (the third Thursday in November) and National Employee Health and Fitness Day (the second Wednesday in May), will enhance marketing efforts. It also involves the company in nationwide movements that are designed to heighten our country's health consciousness.

Work-Load Cycles.   There are cycles within each company that will affect program participation. For example, when companies are preparing budgets or taxes, employees may be required to work longer hours. During these times, many staff members will be too busy to take advantage of the regular programming, and shorter programs

January

- New Year's resolution support program
- Goal setting incentive program
- Health fair
- Health screenings
- Weight control contest
- Winter newsletter and events calendar

National Eye Care Month

February

- Blood pressure screening
- Healthy heart seminars
- CPR courses
- Low-fat menus in cafeteria
- Special Valentine's Day events
- Sweet Hearts Smoking Cessation Program

National Heart Month

March

- Weight control course
- Featured menus in cafeteria
- Nutrition communications program
- Nutrition seminars
- Ethnic restaurant tour/tips
- Shopping tour/education

National Nutrition Month

April

- Cancer screenings
- Cancer prevention seminars
- Summer fitness preparation
- First-quarter health check (progress evaluation)
- High-fiber/low-fat menus in cafeteria
- Cancer awareness communication program
- Spring newsletter and events calendar

National Cancer Control Month

May

- National Employee Fitness
- Day fun run and health fair
- Fitness testing and exercise prescriptions
- Fitness contests
- Walking club
- Fitness seminars

National Physical Fitness and Sports Month

June

- Calcium and osteoporosis seminar
- Fitness during vacation communications
- Summer survival—kids and health
- Sun and skin health
- Relaxation seminar
- Massage breaks

National Dairy Month

July

- Water safety
- CPR refresher course
- Eating right on vacation
- Drinking water for good health and fitness
- Water quality
- Special Fourth of July event
- Summer newsletter and events calendar

Fourth of July Celebration

August

- Annual employee picnic
- Family fitness
- Health preparation for back to school
- Last summer fling fitness outing
- 2nd-quarter health check

Labor Day

September

- Healthy heart fair
- Low-fat cooking demonstration
- Cholesterol screening
- Cholesterol lowering programs for high risk individuals
- Fall program kick-off event
- Fall newsletter and events calendar

National Cholesterol Education Month

October

- Medical consumerism seminars
- Communications on food and drug interactions
- Stress management courses
- Humor workshop
- Grand opening of quiet room
- Halloween celebration
- Communications on healthy Halloween treats

Talk About Prescriptions Month

November

- Smoking risk assessments
- Smoking cessation program
- No-smoking policy implementation
- Great American Smoke-Out Support System (Adopt-a-Smoker program)
- Holiday weight challenge
- Cold-weather exercise seminar

Great American Smoke-Out

December

- Ski-season fitness
- Year in review
- Annual awards program results
- Holiday survival strategies
- Healthy hors d'oeuvres demonstration and taste session
- Communications on holiday eating tips

Holidays

Figure 3.1 Sample program calendar.

53

can be designed to help employees relieve the stress and tension of long hours. For example, shoulder and neck massage sessions or stretch breaks could be a welcome reprieve. The program needs to support employees during these demanding work cycles, not compete for their time and attention.

Enrollment and Health Screening

Following the development of the program calendar, enrollment and health-screening procedures should be established. It is important that these activities be coordinated so the employee experiences a smooth entry into the program. Conducting the screening and assessment is very labor intensive; therefore, scheduling screenings becomes important in developing the program calendar. For example, planning a health fair during a screening and assessment cycle would be a major drain on the staff.

The enrollment process should start at the peak of the marketing campaign, when interest in participation is at its highest. The initial marketing campaign should publicize the dates, times, and locations of programs; an explanation of the enrollment procedures; and an overview of the screening and assessment process. The length of enrollment should be proportional to the penetration goals (e.g., % of eligible employees the program plans to accommodate).

Registration.   The most important principle in developing an enrollment process is to keep it short and simple, because this will be the employees' first exposure to the program. The enrollment process should continue to build the momentum started during the marketing campaign. Many health/fitness programs create lengthy registration packets that contain multiple forms and surveys that confuse and overwhelm participants. The enrollment process will make a first, and sometimes lasting, impression, so it should be a well-organized, pleasant introductory experience for the participant.

Registration forms should be easy to complete and require only the information that is critical for entrance into the program. For a broad and deep program mix, it is more efficient to adopt a universal registration form that can be used for enrollment in most of the program options. In large programs, computerized registration forms can be used to ease the record-keeping chores. Small programs may not need computer sophistication, but they should still consider a universal enrollment process that provides easy access to program delivery. Although the program mix will dictate the specific information required, a comprehensive registration packet should include the following basic information:

- Name
- Work phone/address
- Home phone/address
- Demographics
- Brief health history
- Emergency contact person and phone number

Some programs may have to establish a waiting list because of limited space. One approach is to establish a first-come, first-served policy. Another approach would be to hold a drawing to determine the order of enrollment, providing everyone an equal opportunity to be selected early. It is important to establish the policy at the onset and to publish it in the promotional literature.

Risk Classification.   Developing a process that is time efficient for both the staff and the employees, yet is still adequate to assess risk factors, is a challenge to program planners. First of all, it is important that the screening techniques be medically warranted and that an authorized health professional be available to administer the tests. For example, only physicians can diagnose diseases, and some states prohibit blood drawing by anyone other than a licensed phlebotomist. In New York, California, Ohio, and possibly other states, discussion of blood pressure results constitutes medical practice and, consequently, blood pressure screening by allied health professionals is prohibited. The screening protocols should be reviewed by a physician or nurse practitioner for safety, efficacy, and liability.

Studies have shown that health risk assessment instruments are a cost-effective general health-screening technique.[31] The health risk assessment is a tool that compares an individual's health-related behaviors and characteristics to mortality and epidemiologic data. This data is used to estimate the potential risk of dying by some specified future time. Most of the health risk assessment tools on the market are derived from the data base developed by the U.S. Public Health Service's Center for Disease Control (CDC). Objective measures, like laboratory tests, improve the accuracy of these instruments, but also increase the cost.

In determining which screening tools to use, it is also important to consider the purpose of the screening, as well as the cost-effectiveness and overall practicality of the various devices available. If the intervention presents limited risk to participants (e.g., smoking cessation), it is not cost-effective or necessary to require an extensive screening process prior to participation. In this case, the benefits far outweigh any risks, and the screening procedure can include a brief health history questionnaire that familiarizes the instructor with any individual needs.

The screening process for determining cardiovascular risk prior to exercise participation needs to be more extensive because of the liability associated with exercise programming. The American College of Sports Medicine (ACSM) has published criteria for determining exercise risks and the level of testing required for specific populations.[32] Should there be an exercise-related accident, a company might need to demonstrate that these guidelines were followed. The flowchart in Figure 3.2, based on the ACSM guidelines, diagrams the process for screening employees prior to participation in a fitness program and managing those employees who have been identified as having a high risk of cardiovascular disease.

In some cases, a physician might be hired on a contract basis to review the exercise risks identified by the staff. In a company with a medical department, the company physician can review the risks. Employees could also be referred to their own physicians, who can sign a referral form indicating clearance for exercise. Figure 3.2 depicts the referral loop for medical clearance. Note that employees who are not cleared for an exercise program can still participate in health education programs. It is important for the staff to pay special attention to these employees so they do not feel alienated from the program and develop a negative perception. If the exercise screening process

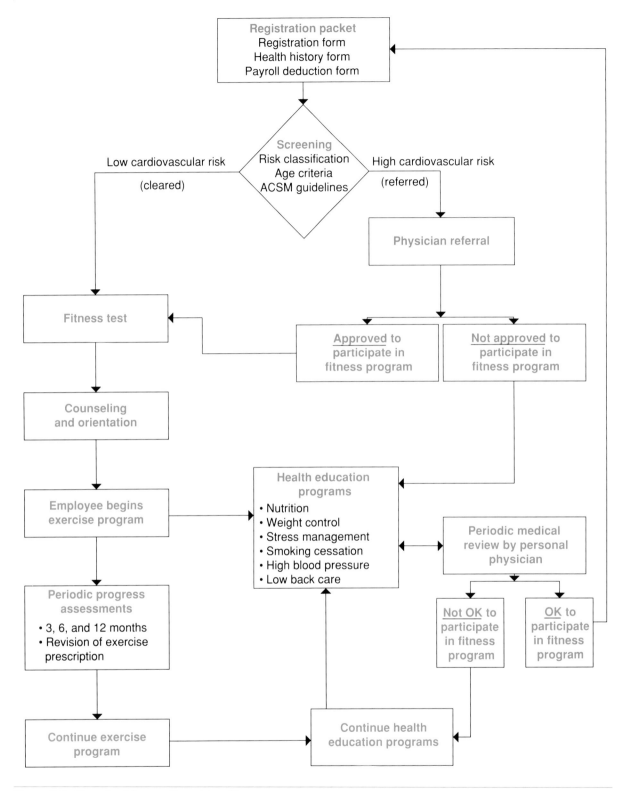

Figure 3.2 Flowchart for enrollment in a fitness program.

is the first time that these employees become aware that they have serious cardiovascular problems, the staff should be sensitive to the fears and other emotions the employees may experience. Establishing a good relationship with the referring physicians is also important in managing these cases, because after a period of treatment, a physician may clear an employee for participation in the program. If the staff have been supportive during this waiting period, both the employee and the physician will be comfortable with the program.

In educational programs, an in-depth screening process is not required for liability purposes but might be used as an enhancement of the program mix. For example, cholesterol screening could supplement a weight loss program by personalizing the information related to decreasing fat consumption. Many of the health screenings and assessments can be used to gather baseline data for comparative purposes in evaluating program results. If the evaluation plan includes these objective measures of program results, the enrollment process should incorporate this data collection stage. Figure 3.3 presents the enrollment and screening process for health education programs. Note that it is not necessary to treat high-risk employees differently from low-risk groups when they are only going to participate in educational programs. Certainly the individual counseling session should address any risks that have been identified, and the employee should be encouraged to contact his or her physician to explore them further. A letter could be provided for the high-risk employees to send to their physicians, but it is not necessary to have physician clearance. If educational programs are being offered without any accompanying screening or assessment activities, the employees can move directly from enrollment to participation.

Health/Fitness Assessments

Every program should offer the employee a way to measure success or failure through pre- and postprogram assessment. Although awareness and education programs do not require the elaborate screening and testing protocols that are necessary for a fitness program, questionnaires help employees understand that they have gained knowledge and made behavior change. Long-term maintenance of a positive health behavior is only possible through continued monitoring that provides adequate and motivating feedback.

Any preprogram assessments that are included in the program mix (Figure 2.2, p. 23) should be scheduled as soon as the employee completes the risk factor screening process. Programs that offer exercise activities usually include testing that evaluates physical performance to identify those individuals who need special attention or medical referral (Figure 3.2). A fitness assessment typically includes the following:

- Aerobic capacity
- Flexibility
- Muscular strength
- Muscular endurance
- Body composition

There are numerous texts that describe in detail the protocols for these fitness assessments.[33-36] Fitness retesting standards will depend primarily on the staffing levels with respect to the size of the participant group. Most corporate fitness centers set annual retesting goals of 25% to 30% of their membership. High-risk individuals should be retested every year, and individuals who do not participate in the exercise program for a period of 12 months should be required to obtain another medical clearance.

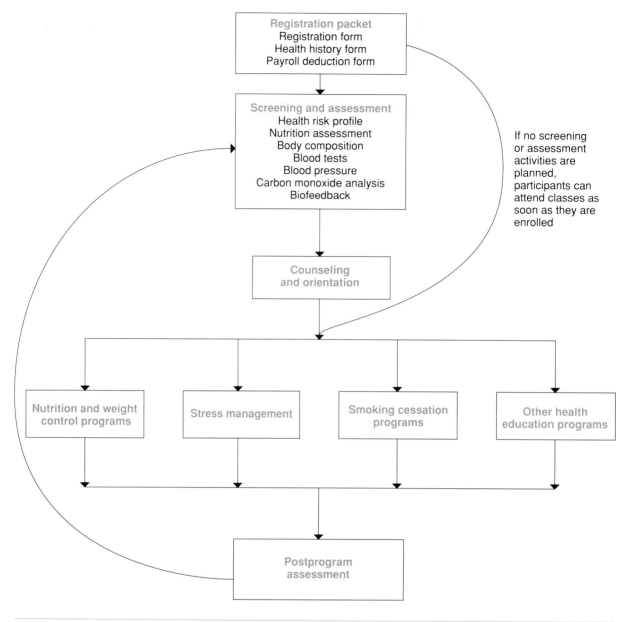

Figure 3.3   Flowchart for enrollment in an educational program.

Health and fitness assessments can be a successful part of the program, but sometimes the assessment and screening process is too labor intensive, requiring staff to spend excessive amounts of time testing and not enough time coaching and counseling employees to implement behavior changes. A major goal in all programs should be to provide employees the skills necessary to measure and monitor their own progress. This is a core step in helping employees learn to take control of their health.

Counseling and Orientation

In any type of wellness program, but particularly in a fitness center, orientation and individual counseling are critical so that employees learn to make changes and remain motivated to stay in the program.

After screening, assessment, and medical approval, the employee should receive feedback on the results of the tests through a small group review or individual counseling session. This can be an opportunity to build a relationship between the staff and employees, allowing for personalized feedback and group interaction.

During a counseling and orientation session the staff should review all of the tests and questionnaires the employee completed. This is best accomplished with a summary booklet (or packet) that reports the results, describes the purpose of the test, explains the normal ranges, and includes guidelines for appropriate lifestyle changes. After reviewing the test results, the staff member can assist the employee in setting goals and developing a plan for change. An initial counseling session for nutrition, weight control, smoking cessation, exercise, and any other lifestyle change should include a behavioral prescription. The behavioral prescription addresses any personal barriers to a lifestyle change. Studies on exercise adherence have shown that it is important to review these barriers when starting adults on an exercise program.[37] Table 3.2 presents common barriers to exercise adherence that should be addressed on an individual basis to help employees understand the level of commitment necessary to be successful.[38]

The last step is to provide an overview of what the participant can expect from the program. In a fitness center setting, the orientation should include a tour of the facility and personalized instruction on how to use each piece of exercise equipment involved in the participant's exercise plan. The counseling and orientation session can take from 60 to 90 minutes. In small programs with limited staffing resources, it may be difficult to schedule individual counseling sessions with each employee. A more efficient approach is to hold group review sessions where each employee receives an individual summary of her or his test results and a staff member guides the group through

Table 3.2   Elements of a Behavioral Prescription for Exercise Adherence

Participant's characteristic	Question
Support group	Does the participant have a person or group that will support his or her exercise efforts?
Skill level	Does the participant have the skills necessary to perform the exercise he or she has chosen?
Convenience	Will this worksite facility be convenient for her or his exercise?
Injury	Does the participant have any past injuries that might stop him or her from being successful?
Satisfaction	Has the participant enjoyed this activity in the past? If so, why?
Motivation	What is the prime motivation to start and maintain an exercise habit?

Note. From "Adhering to Fitness in the Corporate Setting" by M. Landgreen and W.B. Baun, 1984, *Corporate Commentary*, 1(2), p. 39. Adapted by permission of Washington Business Group on Health.

an interpretation of the material. Minimally, each participant should receive an orientation packet that summarizes the results of the assessments and clearly outlines the next steps for the program.

Retention and Motivation

The experiences during the first year of a health promotion program will set the tone for future success or failure. Positive programming experiences early in the program build momentum. Many employees will initially participate in program activities because they are curious. Events that are not well organized or do not deliver what was promised will undermine the employees' confidence in the program and staff. It is important to realize that program impressions left by one poorly organized and delivered program can impact the participation of many employees. This situation can escalate if dissatisfied employees not only quit, but also speak badly about the program.

Once the employees start the program, it is important to continue to motivate them through instruction and personal attention. In addition to a behavioral prescription during counseling, various strategies can be incorporated into the programming plan to increase retention. These strategies include goal setting, an incentive system, and social support.

Goal Setting. Goal setting has been shown to be an effective technique to promote behavior change. During counseling, the staff can work with employees to set short-term and long-term health improvement goals. The goals should be worded in terms that are specific, measurable, realistic, and personalized to the employee's lifestyle. To strengthen the potential for goal achievement, it is helpful to incorporate some of the following strategies into a goal-setting system.[39]

- External monitoring of goal attainment
- Self-monitoring
- Public statement of goals
- Incentives (rewards) tied to goal attainment
- Specific time frame

When the employee sets a goal, the staff should periodically monitor the employee's progress to recognize success and provide coaching for problems. Goal setting can be incorporated into group programs, such as lifestyle courses, to personalize the program for each participant and provide a structure for planning behavior changes. Sharing goals and goal attainment with the group is a powerful way to strengthen the goal-setting process through public declaration and recognition.

Incentive System. An incentive system should balance extrinsic and intrinsic rewards.[40] Extrinsic rewards are obtained externally. This positive reinforcement can come from the participant or from someone else. For example, many programs give participants a small gift (e.g., a T-shirt, fitness bag, or headband) at the beginning to help them feel ownership in the program. Others tie these gifts to attendance or goal achievement. In addition to gifts, rewards can be intangible (e.g., praise, recognition). Simple extrinsic rewards that can be incorporated into the program delivery to promote retention include the following:

Communication

- Postcard of thanks for coming
- Absentee letter sent to participants who miss class(es)
- Weekly phone calls
- Verbal recognition

Incentives

- Attendance gifts
- Rotating award to the most inspirational participant
- Prizes
- Certificate of completion or achievement

Intrinsic rewards are obtained internally (e.g., feelings of satisfaction, self-worth, and fulfillment). In the early stages of a program, participants are more dependent on extrinsic rewards, but the program should wean participants from extrinsic rewards and help them develop an intrinsic motivational structure. These structures are developed by helping participants understand the successes they accomplish and how they feel about these experiences. Intrinsic motivation will be a critical survival strategy during a lapse (a slip or mistake) or relapse (backsliding or many slips added together). Most individuals will experience these slips at some time during the program, so an intrinsic motivational structure is essential for long-term success.[41]

Social Support System. Social learning theories have documented the powerful influence of role models in motivating behavior change.[42] Health/fitness professionals are important role models in a wellness program. Likewise, other employees and family members influence the behavior change process both positively and negatively. As the people in a worksite social system bond through a common interest in health and fitness, more resistant groups are influenced by the program. Strategies to enhance the existing support systems at the worksite (and at home) should be incorporated into the programming activities. For example, buddy systems that encourage employees to support each other by accounting to each other and becoming exercise partners will keep some employees from dropping out. The family unit can be a powerful support system, yet few worksite wellness programs include the family members.[43,44] If family members are excluded by the eligibility criteria, they can still be included in the buddy system.

It is important for programmers to remember that social support positively and negatively influences how individuals cope during a behavior change experience. Good programming staff will teach employees how to be effective in supporting their co-workers. Six important support skills have been identified in Table 3.3. Instruction on these skills can be blended into lifestyle change classes.

The timing of support is also very important. Individuals who are initiating a lifestyle change need more frequent and intense support than a person who is at the maintenance stage. Training in how to seek or ask for support is also useful to some participants. Most employees probably have existing support systems, but fail to use them

Table 3.3 Forms of Support

Form of support	Example
Emotional and psychological	Talking or listening
Instrumental or assistance	Family changes to more nutritional meals
Informational	Sharing information
Motivational	Encouragement
Status	Giving evidence of someone's worth or competence
Companionship	Enjoyment of social relationship

Note. From *Relapse Prevention: Maintenance Strategies in the Treatment of Addictive Behaviors* by G.A. Marlatt and J.R. Gordon, 1985, New York: The Guilford Press.

effectively when making health behavior changes; therefore, adapting old and building new support systems enhances long-term success.

Finally, the corporate culture itself exerts a powerful influence on individual employees' behavior. A culture that embraces wellness will support and motivate healthy lifestyles. On the other hand, corporate cultures that reflect an apathetic attitude (e.g., by permitting smoking) will undermine wellness momentum. Within the corporate structure, the members of the management team are the most significant role models; their health behaviors influence employees' lifestyles in both subtle and overt ways. For this reason, building support and participation within the management group is a critical programming strategy.[45]

Marketing and Promotions

Successful marketing involves more than simply informing employees about program offerings. When a health promotion program is unveiled, many employees will feel relatively healthy and perceive little need to change their lifestyles; therefore, they will not be interested in joining. In addition, factors such as inconvenience, competing demands for time, the psychological and physical discomforts that accompany change, and laziness are frequent barriers to participation. To overcome these barriers, a health promotion program should utilize a wide variety of promotional and merchandising techniques.[46]

In applying these consumer marketing concepts to an employee health promotion program, the marketing campaign should accomplish the following:

- Create an awareness for the program, its mission, and the various ways in which employees may participate.
- Motivate employees to enroll and participate in screening and assessment activities and special events, such as a health fair.
- Sustain participation through the use of ongoing marketing strategies.

Target market analysis and monitoring are essential for achieving all three steps. As discussed in the Marketing Strategy section of Phase

II (p. 30), identifying and researching the characteristics of the total employee population and employee subgroups is critical to developing a marketing plan. Once the plan is implemented, continued monitoring of participation levels, behavior changes, and program interests will allow programmers to measure the success of the marketing campaign. The next section discusses strategies for launching a marketing campaign, as well as how to monitor and evaluate the campaign's success through ongoing market analysis.

Launching a Marketing Campaign

Determining the date for the program launch is the first step in activating the marketing plan. The initial marketing efforts should be carefully planned to coincide with any positive events (e.g., moving into a new facility, start of a new fiscal year) that will enhance excitement about and awareness of the program. Likewise, the launch should avoid any affiliation with negative events (e.g., a layoff). Once the date is established, the program planners should initiate a series of tightly scheduled announcements and awareness-building activities designed to capture attention and stimulate curiosity. Launching these activities in a synchronized fashion requires a team of people to distribute the work load. The employee committee can be assigned marketing roles that suit each person's position and personality. In addition, top management should play a key role if the company expects the program to be recognized as an important initiative. Table 3.4 presents a matrix of responsibilities for program marketing to assist in developing roles for the top management, health/fitness professionals, and employee committee members.

Role of Top Management.   A highly effective way to announce the program is to have the CEO write a letter that is distributed to each employee describing the management's vision for the program and endorsing its inception. Memos from vice presidents and department heads to the employees in their work groups will reinforce the CEO's message and provide specific guidelines for how the program will interface with the department's work. They may have input into the image portrayed in brochures and other printed materials, and they can approve announcements that are mailed with paychecks (check stuffers). Brief presentations during regularly scheduled staff meetings are effective in communicating the program specifics; managers' prior approval is usually necessary to be included on these meeting agendas. During any kickoff or special event, employees will be more impressed and likely to participate if they observe that top management is playing a role. If a keynote speaker is invited, the CEO or another top manager can introduce the speaker and use this time to make a few statements on behalf of the program. The management should play a high-profile role during these awareness-building events. As mentioned previously, health/fitness programming might be incorporated into executive retreats to help top managers develop their own healthy lifestyles and leadership roles.

Role of the Health/Fitness Professionals.   Whether the health/fitness professionals are full-time employees or hired on a contract basis, they can play a critical role in launching the marketing program. The health/fitness professional, as the expert, should have the skills and knowledge to participate in most of the marketing activities. If the

Table 3.4   Marketing and Promotions Matrix of Responsibilities

Marketing effort	Top management	Health/ fitness professional	Employee committee
Communications			
Letter from CEO	x		
Posters/fliers		x	x
Memos	x	x	x
Brochure	x	x	x
Check stuffers	x	x	x
Presentations/meetings			
To employees	x	x	
To management	x	x	
Meetings with supervisors	x	x	
Special events			
Kickoff event	x	x	x
Health fair		x	x
Open house		x	x
Keynote speaker		x	
Management health retreat	x	x	
Employee referral			
Incentive program		x	x
Word of mouth			x
Program representative circulating work areas		x	x

staffing model involves hiring a full-time professional, this person should initiate the marketing activities and coordinate the involvement of management and the employee committee. If the company has planned to hire contract professionals to staff the program on a part-time basis, the program coordination should be assigned to an employee who can seek advice from the consulting health/fitness professionals. In this situation, the contract professionals can assist in designing and writing the communications, conduct presentations, meet with supervisors, plan and attend special events, and oversee the employee referral network. The internal coordinator, if she or he has the time and skills, could manage this work. When health/fitness professionals play a role in launching a marketing program, they can be positioned as healthy role models and experts who inspire the employees to join the program.

Role of the Employee Committee.   The employee committee will be helpful as a pilot group to assess how employees will react to the materials and strategies. When the marketing materials are completed, the committee members can assist with distributing and posting the fliers and brochures. Many special events are labor intensive, and committee members can volunteer to assist with decorations, equipment setup, crowd control, enrollment, record keeping, and cleanup. Members who do not assume specific roles can network with other employees, promoting the program and answering questions. Committee members will play an important role in the referral program through word-of-mouth promotion within their work groups. An incentive program of awards and prizes for employees who enroll their friends and co-workers stimulates peer pressure and rewards them for promoting the program. Within the total marketing campaign, the employee committee becomes the vehicle for reaching the grass roots.

Market Analysis and Monitoring

Although the initial market research will come from the needs analysis, ongoing market analysis will result from the program records. Within every health promotion program, participation rates should be tracked and the results summarized on a regular basis (e.g., monthly or quarterly). In a small company, this task can be delegated to the champion or the internally assigned program coordinator. At an absolute minimum, the program director or coordinator needs to know (a) the number of participants enrolled, screened, and assessed, and (b) the attendance or participation rates of each program offering. Record-keeping systems are discussed in more depth in the Operations and Administration section (p. 68). With this information, the results can be compared to other programs or industry norms to determine strengths and weaknesses in the program design and marketing efforts.

For large groups (e.g., over 1,000 eligible employees), the analysis should be performed on subgroups to identify underserved or poorly served segments (e.g., night-shift workers, sales staff) within the total population. It is important to segment the market into demographic, psychographic, and behavioristic bases, because each type of analysis will reveal different characteristics and patterns.[47] Demographic bases refer to criteria such as age, gender, family size, income, and occupation. Psychographic bases are things that have a direct influence on a person's lifestyle, such as needs, interests, attitudes, and values. Behavioristic bases define the behavioral characteristics of target groups and include such information as commuting patterns, weekend recreation, and shopping and cooking habits. An annual survey to determine overall awareness and satisfaction will facilitate this analysis.

No two programs are alike, so comparison of the results between programs is complicated. Nonetheless, Table 3.5 offers a rough guideline for comparing participation rates in a fitness center for 500 to 2,000 eligible employees with at least 3 square feet per employee. Even within this standard, it should be noted that the smaller programs tend to have higher participation rates, assuming that other factors are equal (i.e., staffing levels, facility availability, and employee copayments).

An annual survey to determine the overall awareness and satisfaction will facilitate the market analysis. Constructing and administering an annual survey will be discussed in the Quality Assurance section (p. 78). Without the benefit of a survey, program directors can use enrollment rates to gain insight into the level of awareness about the program. If a program has less than 35% enrollment, heightening awareness through a marketing blitz might remedy the problem.

Table 3.5   Average Penetration Rates Achieved at Specific Marketing Stages

Marketing step	Percentage of eligible employees	Percentage of employees enrolled
Aware of program	80-90	—
Enrolled	35-65	100
Frequent participation	20-50	50-75

Note. Rates based on established programs with 500 to 2,000 eligible and at least 3 square feet of facility per eligible, day-shift employee. From proprietary data from Fitness Systems, 1989.

Oftentimes this analysis will reveal that the problem goes back to program basics (e.g., staffing ratios, quality of program, convenience of program offerings, program mix, top management support) that marketing techniques cannot overcome. These are critical factors to recognize, because it is impossible to resolve a problem when its true cause is not clear.

Staff Selection

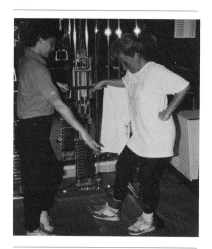

Professional staff can be the most important element in the successful operation of a health/fitness program. Staff ensure that the program activities are safe and based on sound scientific principles. They also serve as role models of health and fitness, motivating employees to adopt lifestyle changes. Qualified staff increase the opportunity for individual success by providing employees with up-to-date information and programming activities. The staffing model developed in Phase II, Conceptual Definition (p. 32), will guide the selection of professional staff, although it may be necessary to revise the staffing model to fit the candidates available when it is time to hire. Recruiting, hiring, training, and supervising the professional staff according to the previously defined staffing model should occur during the implementation phase. Critical to the selection of high-caliber staff is an understanding of the entry-level and advanced qualifications for professionals in the health/fitness industry.

Qualifications and Certifications

The minimum requirement of an undergraduate degree in exercise science, health education, physical education, nutrition and dietetics, nursing, or a related health science has become a standard in the health/fitness field. As the complexity of positions increases, the need for graduate training arises. A master's degree is common for the health/fitness professional who desires career advancement. Business skills are becoming increasingly important to health/fitness professionals practicing in corporate settings or management positions. The program-director-level position should require a combination of training in a technical field and practical experiences in management roles.

In addition to the formal coursework available in degree programs, there are a number of certification programs that impart specialized skills to the health/fitness professional. Appendix D presents a resource list of organizations involved in certifying health/fitness professionals. The Aerobics and Fitness Association of America (AFAA) and the American Council on Exercise (formerly the IDEA Foundation) both offer certifications to professionals involved in teaching group exercise classes (aerobic dance). The YMCA has an excellent series of certifications, restricted to YMCA employees, with instruction and competency testing in a wide range of health/fitness skill areas.

Currently, the strongest and most recognized certification program is offered by the American College of Sports Medicine (ACSM). Certification is divided into health/fitness and rehabilitative tracks. The rehabilitative track is designed for professionals involved in a hospital or rehabilitative environment and involves three tiers of competency or levels of certification: exercise test technologist, exercise specialist, and exercise program director. The health/fitness track is designed for professionals who will be working with asymptomatic individuals and also includes three competency levels: exercise leader, health/fitness instructor, and health/fitness director.[32] The health/fitness track certifications are most applicable to professionals seeking positions in corporate settings. However, certifications, like educational degrees, are only markers of potential knowledge gained; it is important that this information be used in combination with the candidates' practical work experience and references to evaluate their level of professional development.

Recruitment

The procedures for recruiting and hiring staff will be based on the staffing model (i.e., hiring full-time staff or contracting with an external provider organization), as well as any internal personnel policies. If a full-time professional will be on the company's payroll, the program director is hired before the support staff positions are filled. The program director can then manage the recruiting and hiring of other staff. If an external provider organization is being retained to staff the program, the company will need to spend time soliciting proposals and meeting with potential providers before making a selection. Appendix D provides names and addresses of recruiting networks and directories of provider organizations.

This first hire (whether it is a full-time professional or an external provider) is a critical decision for the company and requires time and attention from several individuals to assure that a good decision is made. It could take years to overcome the negative momentum resulting from a poor staffing decision at the onset. This process will need to be managed at a high level, but the employee committee can play a role. When they have an opportunity to be a part of the recruitment process, their ownership for the program and future staff increase. Committee members can assist during interviews by giving candidates tours of the corporate facilities and an overview of the program.

Patton et al. suggest a four-step process that allows both the company and candidates to evaluate each other's desirability for a given position: advertise, evaluate resumes, check references, and interview.[29]

Internship Programs

Expansion of health/fitness programs has encouraged the development of various undergraduate and graduate programs in colleges and universities. Most of these programs require students to complete a successful internship in an established worksite or hospital setting. The primary goal of an internship program is to provide the student an opportunity to gain practical experiences and skills that are required in the role of a health/fitness professional. These programs are organized so that the student, academic institution, and collaborating company all benefit from the experience.

To initiate an internship program within a company, the first step is to obtain approval from upper management. The duties and responsibilities of interns need to be determined in order to assess the time demands of the supervisor, potential costs, and office space requirements. Some corporate internship sites develop a brochure or packet that describes intern qualifications, responsibilities, length of internship, and financial arrangements. This material also describes the program, staff, activities, and facilities and includes a brief statement about the company. Prospective interns will use this information to determine which program will allow them to accomplish their individual academic goals. An application form should be included in the packet.

Solicitation of qualified intern candidates usually starts by contacting faculty members of universities and colleges that have programs in health and fitness. It is also helpful to publicize the program in AFB's Internship Clearinghouse (see Appendix D). Interested students should be asked to submit resumes, letters of recommendation, and a statement outlining their expectations and goals. A screening process should be established to review applications and select interns. It is important to develop an understanding of each student's career objectives in order to provide meaningful experiences during the internship. Student interns can be an inexpensive labor pool, but supervisors of internship programs have a responsibility to the student and sponsoring institution to make sure that the student learns and grows. Sometimes the time demands may be too great for the supervisory staff.

Operations and Administration

Effective operations affect the long-term management of the program. Operational problems with enrollment, record keeping, equipment breakdowns, untidy facilities, and unpleasant or incompetent staff result in high attrition rates and a bad reputation for the program. Even though proper documentation and monitoring of operational procedures is essential for consistent and high-quality programming, it is frequently overlooked. A policy and procedure manual that documents the program and facility operations helps to establish consistency in program delivery. A facility cleaning and maintenance routine will keep a facility in top condition. Proper record keeping assures that data is collected and maintained in a system that is easy to analyze and report.

Policy and Procedure Manual

The policy and procedure manual addresses key administrative issues involved in the operation of the program. The manual can be organized in a number of different ways and could be divided into several different manuals, depending on the scope of the program. It is important that these manuals provide the staff with a comprehensive set of documents that will increase their effectiveness in safely managing the day-to-day operations of the program. They should be well-organized into logical sections, separated by tabs, and include a table of contents and an index. Appendix G presents a sample table of contents, divided into appropriate sections, for a manual that would guide the administration of a comprehensive fitness program and facility.

Such documentation is obviously essential for large programs in training new staff members and maintaining continuity in the face of inevitable staff turnover. However, even small programs will need guidelines for substitute staff coverage during a program director's absence. A new staff member should be able to read the manual and gain an understanding of the daily program administration. Each staff member should maintain a copy for continuous reference. Because a goal of programming is to constantly improve the program, these manuals should be working documents that can be easily amended; in fact, they should be reviewed and updated annually.

Facility and Equipment Maintenance

If the program mix incorporates a fitness center into the operations, it will be important to establish cleaning and maintenance schedules to prolong the life of equipment, meet sanitation standards, and project a spotless image. Facility maintenance is a constant challenge to a fitness director. Showers that are growing mold or equipment that is continually breaking down because of poor maintenance can very quickly have a negative effect on participation.

The first step in establishing a maintenance strategy is prevention. Preventive maintenance starts with building a facility using high-quality materials that can be cleaned easily. It also starts with purchasing equipment that not only has good maintenance records but can be serviced locally with vendors that promise short response times. A quality maintenance program involves the development of a plan that, first, identifies the maintenance requirements and, second, provides a strategy for fulfilling these requirements. A detailed review of the facility will result in a list of the areas and equipment that must be maintained, as well as the specific needs of each item on the list.

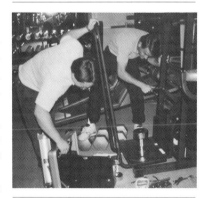

This checklist, or cleaning and maintenance schedule, will become the monitoring tool to guarantee that the specifications are followed. National, state, and city health ordinances will dictate cleaning and maintenance standards in some areas of the facility, such as showers and bathrooms. Owner's manuals and warranty packets will provide special instructions for each piece of equipment. The section in the policy and procedure manual on cleaning and maintenance should include a list of the tasks to be completed, the frequency at which each task is to be performed (e.g., daily, weekly, monthly, after each use), materials required, completion standards, and the estimated time to complete. The maintenance duties need to be scheduled into the staff's work load to make sure that maintenance is not overlooked when the staff is busy conducting other activities.

In addition to the regular cleaning and maintenance schedule, an equipment service company can be contracted to provide more technical support. They can visit the facility periodically to tighten belts on the treadmills, check the tension on the bicycles, lubricate motors, and anticipate breakdowns. This enhances the lifespan of the equipment and prevents expensive repairs later.

The most important component of a preventive maintenance program is ensuring that participants have learned how to operate the equipment correctly and help by wiping off their sweat after using the equipment. It is also important to provide participants a means to communicate maintenance problems to staff—for example, via a facility suggestion box. Keeping a fitness facility immaculate and in good working order is part of what distinguishes a quality operation.

Record Keeping

Many different forms will be used in the administration of the program. Copies of the forms should be maintained in an appendix of the policy and procedure manual, and instructions on how to complete these forms and maintain accurate records should be discussed in the manual. Record keeping is an important responsibility of the staff, but it is also important that the systems are not so cumbersome that the staff is bogged down with paperwork.

Forms and Protocols. Table 3.6 presents an exhaustive list of the common forms and protocols used in a health/fitness program. The enrollment, testing, and participant forms are useful for documenting the status of each participant (e.g., payment, physician clearance, and progress) and should be maintained in individual files for each employee. The administrative forms are useful to the day-to-day operation of the program or facility, covering such areas as staff management, facility management, emergencies, financial records, and quality assurance.

A large, comprehensive program that includes a fitness center might require all of these forms; a small, education-oriented program might only need the following forms and records:

- Registration form
- Health history form
- Fitness assessment forms
- Participant records
- Behavior change logs
- Attendance records
- Body weight records
- Goal-setting records
- Food diaries
- Instructor performance
- Program results summary
- Employee satisfaction survey

An orderly filing system will be required to maintain proper records. Typically, files are organized into participant files, financial

Table 3.6 Common Forms and Protocols

Program operations	Administration
Enrollment/screening	**Facility management**
Health history form	Supply inventory form
Physician release	Inventory records
Registration form	Maintenance checklist
Participant waiver form	Daily facility maintenance log
Locker registration form	Equipment sign-out form
Guest participation form	Library check-out form
Testing and teaching	Court sign-up sheet
Informed consent	Lost-and-found form
Fitness assessment form	Suggestion sheets/box
Strength testing form	**Staff management**
Participant records	Time sheets
Behavior change logs	Schedule sheets
Attendance records	Performance review form
Body weight records	Training request form
Exercise logs	**Emergency procedures**
Goal-setting records	Injury report form
Food diaries	Accident report form
Cigarette tally sheets	**Financial records**
	Budget estimate forms
	Copier use log
	Long distance phone log
	Signature authority form
	Quality assurance forms
	Instructor performance
	Participant evaluation
	Employee satisfaction survey
	Program results summary
	Postprogram follow-up

records, personnel records, and general administrative files. Some of these records can be maintained manually; others are more effectively maintained by computer. These choices will depend primarily on the size of the program and evaluation plan, as well as the availability of computer hardware and software.

Computers. For both large and small programs, computers are important to record keeping and communication functions. The large quantity of information required to administer a health promotion program demands the use of a computer system to maintain records and access data. With the relatively low cost of data-base and spreadsheet software, as well as the availability of specially designed software for the health/fitness industry, it is mostly a matter of determining what computer system to use, rather than whether or not to install one.

In a fitness center, commercially available exercise software could be installed. A computer terminal positioned at the front desk of a fitness center can be used by members to check in and log their activity. Many of these exercise-logging programs have the capability to calculate caloric expenditure, aerobic points, total miles, and other quantifiable factors, which provides the individual employees feedback on their goal achievement and lifestyle changes. Computers placed in exercise areas can also be used for logging exercise, and one advantage is that they immediately calculate the calories that were burned during the exercise session. At the end of each month, the employee can receive a written report that lists the exercises and calories burned during the month. These monthly reports motivate employees and help them track their progress. The records can also be used as the basis for incentive programs by tying rewards to the achievement of specific goals or competitions.

In programs without fitness facilities, interactive computers set up in reading rooms or libraries can be used as a self-help learning activity. For example, a number of nutrition software programs evaluate the nutritional and caloric value of food and plan healthy menus. A new application of computers in worksite health promotion programs is to provide access to health promotion software at each employee's workstation. Employees can log their exercise progress or look up the calories of their lunch using computers on their desks. This makes program resources conveniently available to many employees, even those who work in remote sites (e.g., branch offices).

Probably the most important use of a computer in a worksite program is its potential for providing feedback to the staff on utilization statistics. Software that tracks participation for each employee provides an efficient system for tracking utilization statistics. Worksites that manage their programs through the use of computers have better evaluation opportunities. Utilization data, demographic data, medical claims, and absentee records can all be merged into one computer data base to facilitate analyses. Without a computer, sophisticated evaluation cannot be performed.[48]

Summary

During Phase III, Implementation, the plans developed during Initial Planning and Conceptual Definition (Phases I and II) are activated. A program calendar is developed, and specific enrollment and screening procedures are established. Any planned assessments should be scheduled to make sure baseline data for evaluation is gathered before the intervention begins. New program participants should be provided with an orientation, whether it is a counseling session or an information packet. Strategies to maximize retention and motivate behavior change should be initiated. Launching the marketing plan is a team effort, involving the top management, health/fitness professional, and employee committee. Once the marketing campaign has been launched, ongoing market analysis helps to continually promote the program and refine marketing strategies. Selecting qualified staff to fill the staffing model developed in Phase II is an important aspect of implementation. Finally, setting up operational systems, such as a policy and procedure manual, a facility and equipment maintenance schedule, and record-keeping systems, ensures smooth program delivery and management in the long run.

PHASE IV

Evaluation

Evaluation
• Project evaluation • Periodic reviews • Longitudinal data analysis • Results interpretation and communication

The systematic evaluation of a program entails project evaluation to quantify program results, periodic reviews of progress and program quality, and longitudinal evaluation reviewing program results and changes over time.[38] As the figure on pages viii and ix depicts, the evaluation phase is an end point and a starting point. The information gained through evaluation can be used to plan better relapse prevention strategies, develop new programs, and recharge old programs so they remain effective. The results from an evaluation reveal important feedback for Phases I, II, and III, enhancing the program's growth and evolution.

Health/fitness professionals who have master's degrees will be familiar with basic concepts of statistics and research methodology; however, a sophisticated evaluation project is beyond the skill level of most program managers. This section provides an overview of evaluation methods that are commonly applied to the evaluation of work-site health promotion programs and discusses some of the specific applications. Table 4.1 presents an outline of the evaluation systems that might be incorporated into an employee health promotion program. Although it may be necessary to hire faculty members at a local university to assist with in-depth program evaluation, program managers can use this overview to determine what types of evaluation to implement.

Project Evaluation

Programmers rarely have an opportunity to work on only one activity at a time. Generally they find themselves in different programming phases on a variety of projects. Consequently, evaluation, the last phase in the programming process, is frequently neglected. Project evaluation is divided into outcome, impact, and process categories.[49]

Outcome Evaluation

Outcome evaluation focuses on the question, Did the program cause the results that occurred? A few examples are listed here:

- Reduction of the average blood pressure following a hypertension reduction program
- Decrease of back pain incidence after a healthy back awareness program
- Lowered weight following a weight reduction program

Table 4.1 Overview of Evaluation Process

Evaluation type	Suggested tools/processes	Questions addressed
Project evaluation	Outcome evaluation	Did the program cause the following results? • Reduced hypertension • Reduced back pain • Decreased absenteeism • Decreased employee turnover • Decreased medical care cost
	Impact evaluation	To what degree did the program affect the following? • Behaviors • Attitudes • Company culture
	Process evaluation	Did program administration affect the following? • Enrollment • Attendance • Skill acquisition Were delivery methods effective? • Population-specific • Delivery modes • Cost-effectiveness
Periodic reviews	Quality assurance	Are high-quality operating standards being maintained? Are employees satisfied with the program? Are the staff performing according to the program standards?
	Monthly reviews	Summarize daily and weekly statistics • Utilization • Penetration • Adherence
	Quarterly and semiannual reviews	Analyze effectiveness of program mix. Review accomplishment of goals and objectives.
	Annual review	Review key results from monthly and quarterly reports. Review accomplishment of goals and objectives. Conduct employee survey.
Longitudinal data analysis	Program tracking	Who is participating in the program? • Age • Gender • Job category Who are the program dropouts? • Age • Gender • Job category
	Behavior change tracking	What lifestyle changes are being made? • Health-screening comparisons • Health-age comparisons
	Cost-benefit analysis	How do the program costs compare with their expected benefits, expressed in monetary units?

Outcome evaluation is useful in determining whether lifestyle programs change health indicators, such as weight, blood pressure, blood cholesterol, and other risk factors. Outcome evaluation is also used to examine the savings realized through the reduction in absenteeism, employee turnover, injury rate, and health care claim utilization. These measures are briefly discussed in the Cost-Benefit Analysis section (p. 82).

Impact Evaluation

Behavioral, attitudinal, and cultural changes that occur following program initiation are measured by impact evaluation. Behavior and attitudinal changes can be measured using surveys to study pre- and postprogram changes. Outcome evaluation documents that the intervention caused specific results to occur, whereas impact evaluation measures the degree of change resulting from the program. An example would be to look at the nutritional habits of employees before program start-up, and then use the same survey at the end of the program. The survey could be used to assess the degree to which employees made nutrition-related behavior changes, and how these changes affected the corporate culture.

A major set of questions that can be answered by impact evaluation relate to how the program affects the corporate culture. According to Allen et al., culture refers to any group of people who get together over a period of time with shared goals and values.[50] The more realistic method to define and measure culture is to define the set of "organizational norms" within the group. Organizational norms are traditions that dictate how things are done or not done. Allen suggests eight crucial areas for identifying norms in any organization:

- Rewards and recognition
- Modeling behavior
- Confrontation
- Communication and information systems
- Interactions and relationships
- Training
- Orientation
- Resources commitment and allocation

All eight areas provide many different norms that can be measured and evaluated upon initiation and at the end of projects. These measurements can be obtained through focus groups or surveys.

Process Evaluation

Process evaluation deals with the qualitative aspects of program delivery, such as program registration and instructor effectiveness. It focuses on how the program works from an administrative perspective. A process evaluation might use a participant satisfaction questionnaire or a focus group to detect problems with the administration of the program. Process evaluation could also be used to compare different types of delivery to determine which was the most effective. By looking at the age, gender, and job category of employees who participated in the various program offerings, the evaluator could assess whether one delivery mode or teaching method appears to draw certain segments of the population. For example, one program might entice participants into a 10-week course by giving them T-shirts after program completion, and another program might simply provide a free drink during a lunchtime health lecture. Process evaluation could be applied to answer the questions, What types of employees are drawn into each of these programs, and what aspect of the program draws them (i.e., the free gift, or the type of program)?

Cost-Effectiveness Analysis

Cost-effectiveness analysis (CEA) helps the program planner determine which program produces the greatest benefit for the lowest cost. CEA adds to the subjective information gained through surveys, interviews, and focus groups. Chenoweth lists four steps in performing a CEA.[51]

1. Determine the program objectives or what the program is designed to do for the employees or company.
2. Determine the total operating costs for each program to be evaluated.
3. Determine the outcome of each program used to meet the specific program goals.
4. Compare the program outcomes, and determine which program is most cost-effective.

Table 4.2 presents an example of CEA used to compare two approaches to exercise program delivery—the Buddy Exercise System and the High Energy Incentive Program. The Buddy Exercise System is a program that requires everyone to participate in pairs, and the High Energy Incentive Program is an incentive program that provides a T-shirt after successful participation. The Buddy System costs $1.25 less per successful participant than the High Energy Incentive Program, but the High Energy Incentive Program had a 20% higher success rate, at only a 10% increase in cost. There was also very little (4%) difference between the number of individuals who were successful in each program. Because of the high success rate, very small difference in cost, and different employee groups that were reached with each program, it makes sense to use both programs again.

Periodic Reviews

Periodic (short-term) evaluations provide an opportunity to monitor the qualitative and quantitative results of a program at regular inter-

Table 4.2 Cost-Effectiveness Comparison of Two Program Strategies

Type of program	Cost of program	Number of participants	Cost per participant	Percentage success (number of successful participants)	Cost per successful participant
Buddy Exercise System	$2199	351	$6.26	55% (193)	$11.39
High-Energy Incentive Program	$2327	245	$9.49	75% (184)	$12.64

Note. From an unpublished management report by Tenneco Health and Fitness Program, 1987, Tenneco Management Company.

vals. Critical to the successful use of periodic evaluation is the selection of a time frame that is meaningful from a measurement and programmatic standpoint. On a daily and weekly basis, quality assurance measures might be performed to maintain certain operating standards. Monthly reviews provide an opportunity to focus on short programmatic segments. These segments over time will be consolidated when reviewing the program's quarterly, biannual, and annual progress. Periodic reviews summarize accomplishments in relation to the original goals. This systematic process helps to keep the staff focused on the program's goals and objectives, and how well these goals are being completed.

Quality Assurance

Quality assurance is the continual process of monitoring and measuring performance against quality standards, in an effort constantly to improve a product or service. A quality assurance system integrated into the overall delivery process maintains superior performance by checking outcomes against standards and taking steps to correct deviations. Because quality assurance deals with the qualitative aspects of how the program is administered, process evaluation methods are frequently used.

Quality Standards. The first step in developing a quality assurance system is to define standards, based on industry norms, against which the outcomes will be measured. These standards will encompass variables related to

- behavior change (e.g., food choices, eating habits, exercise habits),
- health status (e.g., blood pressure, weight, body composition, aerobic capacity, strength), and
- customer satisfaction (e.g., attrition rates, attendance patterns, instructor rating, needs satisfaction).

For every program offering, these predetermined standards should be set. Table 4.3 presents sample quality standards for various program offerings within the program mix. Because hospital accreditation involves extensive quality assurance, most hospital-based health promotion programs have established quality assurance standards to cover all aspects of program delivery. Note that some of the standards included in Table 4.3 are adapted from hospital programs. Some commercially packaged programs have established quality standards through pilot testing and extensive intervention trials, and provide tools for the program manager to implement quality assurance. An internally developed program would need to establish similar quality assurance standards, tailored to the program's goals and objectives.

Quality Assurance Checks. During the program delivery process, quality assurance checks should be implemented to monitor interim and final outcomes. For example, when using skinfold measurements to test body composition, reliability checks should be conducted to reduce the deviation from one tester to another. In educational programs, instructor performance should be evaluated to assess teaching skills, voice quality, body language, program knowledge, and logistics (e.g., scheduling room arrangements, AV equipment). Aerobics instructors should be monitored to make sure that unsafe movements and body mechanics are not being taught.

Table 4.3 Sample Quality Standards

Awareness/education programs

Achieve an 80% awareness level after 6 months of marketing and programming.

Attain an average attendance level of 70% in a seminar series.

Evaluation/screening programs

Screen 70% of the employees in an annual cholesterol testing program.

Maintain a variance among skinfold testers of ±3 mm per site.

All health fairs and screenings will be conducted in accordance with applicable laws and regulations.[a]

Gloves will be worn by all staff involved in finger-stick blood draw procedures. Gloves will be changed after every participant.[a]

Prescriptive programs

Lose an average of 8 to 10 pounds in a 10-week weight loss program.[b]

Decrease an average of 1% body fat for every 2- to 4-pound decrease in body weight.

Staff

All staff are required to successfully complete a basic CPR course on a yearly basis.[c]

Screening staff must complete a training program in testing protocols and pass a practical exam.[a]

All teaching staff must complete a 2-day training program in the content of each course they will teach, and give a sample presentation.[b]

Teaching staff must receive a score of 80% on the Instructor Performance Form before being assigned a class.[b]

Facility standards[c]

No one waits at the reception counter for more than 30 seconds before being noticed.

No equipment will be down for more than 24 hours without an adequate written explanation for the member.

New members receive a welcome letter within 1 month.

Each member is greeted with a smile and is treated with courtesy and respect.

[a]Adapted from the Quality Assurance Standards for Health Screenings from Methodist Hospitals of Memphis, Memphis, TN. [b]Adapted from Total Quality Management Program, ProActive Health, Rush-Presbyterian-St. Luke's Medical Center, Chicago, IL. [c]Adapted from the Quality Assurance Program from Parkside Sport and Fitness Center, Park Ridge, IL.

Immediately postprogram, and at periodic intervals, follow-up results should be tallied and compared against the standards to assure that participants have accomplished the program goals and are maintaining behavior changes. The follow-up data can be collected by individual assessments, written questionnaires, phone surveys, or face-to-face interviews.

On an annual basis, a company-wide survey should be conducted to determine the perception of the program. These results can be used for ongoing market analysis (p. 65), so the survey should be distributed to both participants and nonparticipants. The contents of the survey should address the following:

- Awareness of the program
- Perception of the program quality
- Level of interest in the program offerings
- Health/fitness needs and interests

- Effectiveness of program offerings (e.g., availability, location, enrollment procedures, and hours)
- Behavior changes resulting from the program

Corrective Actions. When quality assurance reveals that certain elements of the program do not meet or exceed the standard, corrective actions need to be identified and implemented. This process ensures that the customers (employees) are satisfied with the service and the program performs uniformly against predetermined standards. For example, if reliability checks on testing procedures reveal that there is a high deviation when different staff members conduct body-fat assessments, a staff development program should be initiated to establish consistency in the testing protocols. If an instructor receives low ratings in class presentations, he or she needs to conduct several mock classes before being assigned to teach again. When a program fails to achieve program standards (e.g., no one loses weight in a weight loss program), the quality of the instructor, program content, time of year, and other variables need to be investigated.

If the annual survey reveals that a high percentage of the employees are not aware of the program offerings, this indicates that there is a marketing problem, and steps should be taken to conduct a target market analysis and launch an awareness-building campaign. If enrollment and attrition are both high, this indicates that awareness is not the problem, but retention strategies are needed. These ongoing steps to evaluate and improve the delivery process will assure that superior performance is consistent and slips in quality are short-lived.

Monthly Review

The monthly review should compile and summarize daily and weekly statistics on utilization, penetration, and adherence. These quality assurance and marketing indicators should be collected routinely to help the program director manage the program. Gathering these statistics on a monthly basis to review the progress provides powerful information to help the staff make more successful programming decisions. The program coordinator or director oversees the monthly review process, distributes the work load and delegates components of the report to other staff members. Once the entire report has been integrated, the staff should review the completed report and forward it to the next layer of management.

Utilization. Program utilization is simply the number of individuals involved in program activities for the month. These counts can be used to identify peak participation periods, attendance cycles, and seasonal variation. This information helps staff develop strategies that increase utilization and identify natural cycles in program enrollment so they can make better scheduling decisions.

Penetration. When utilization is divided by the total population, the quotient is a rate that measures how far the program has penetrated the total population. This measure is called the penetration rate. It is helpful to calculate penetration rate when evaluating the success of targeting selected market segments. Good examples would be measuring the depth at which membership has reached the middle-aged population after a membership campaign for a fitness center, or the number of smokers who sign up for a smoking cessation program. Breaking the population into various market segments by age, gender, and job category allows for the development of target market strategies and interventions. The evaluation of these strategies and interventions through a monthly review process monitors overall employee utilization and program penetration.

Adherence. Program adherence measures the regularity of participation and is an indicator of the program's success. If employees stop attending classes or using a fitness center, they will not benefit from the program. Typically, at least half of the individuals who engage in a new health behavior do not continue with it,[52] and attrition rates in dietary programs may be as high as 86%.[53] Clearly, attaining high levels of adherence in a worksite wellness program is a challenge. Routine monitoring of adherence levels will help staff develop retention strategies.

Quarterly and Semiannual Reviews

Periodic evaluation might also include a quarterly report. At quarterly intervals, the integration and interaction of several program activities can be analyzed, allowing a programmer to review the effectiveness of the program mix. The benefit of quarterly evaluations for most programs is that program calendars are generally created in 3-month segments, allowing the staff to evaluate the promotion, registration, and delivery activities of each programming cycle. Programs that do not use quarterly reviews may complete a semiannual review that provides information about the first half of the programming year. If the program has experienced a slow start, a semiannual analysis will catch problems in time to make up the difference during the second half of the year. These reviews can also be used as markers toward the accomplishment of annual goals and objectives.

Annual Review

The final periodic evaluation is the annual review. Unlike the monthly, quarterly, or biannual review, this analysis also includes some longitudinal variables. The annual review process involves compiling key results from the monthly, quarterly, and semiannual reports and relating the status to the accomplishment of the program goals and objectives. Some programs are managed by the MBO (management by objectives) approach, so this final review is the precursor to the establishment of new goals for the coming year.

As discussed in the Quality Assurance section (p. 78) many programs survey the employee population on an annual basis. This survey is a major component of the annual review. One strategy that will deter low response rates to annual surveys is to survey different market segments each year. It is very important to receive this annual feedback from both participating and nonparticipating employees, representing a cross-segment of the population. Once this information has been collected, staff can address the potential need for making changes in the programming strategies and interventions that will better serve the entire population.

Longitudinal Data Analysis

The last and most complex evaluation category is the longitudinal design. This approach requires the evaluator to track employees over years of program participation and nonparticipation. It is expensive and time consuming, and in a worksite environment, participation dropout due to employee turnover creates a major problem in longitudinal analysis. Many of the evaluation variables that have already been discussed can also be reviewed through longitudinal design. For example, tracking program participation by age, gender, and job classification provides important data concerning the participation shifts that occur in a program over time. Annual screenings or testing programs (e.g., for blood pressure, cholesterol, or fitness level) measure changes in health status over time.

Behavior Change

The ultimate goal is adherence to a healthy lifestyle, thus compliance becomes a major component of longitudinal studies. Collecting health risk assessment data over several years is one of the easiest and least expensive longitudinal measures. It allows for the measurement of a multitude of health behavior variables, and the health-age algorithms provide a good single measure for the overall effect of the program.

Cost-Benefit Analysis

Cost-benefit analysis (CBA) involves a longitudinal data base because these studies require measurements collected over many years. CBA compares the monetary costs of programs with their expected benefits (expressed in monetary units). If the program has been organized as a cost-containment strategy, then CBA will determine the worth of the investment. Although CBA is of major importance to the continued growth of health/fitness programs, the cost effectiveness of completing such a study has been questioned by many evaluators.[54] CBA studies are impractical for most worksite programs because of the cost, the large sample required, and the lack of historical data. Information concerning the steps necessary in completing a CBA can be found in the article by Murphy et al.[55]

Results Interpretation and Communication

An evaluation system can break down for a number of reasons, one of the most significant being the failure to properly interpret and communicate the results to the appropriate individuals. Within an employee health promotion program, there are many different levels of communication important to the success of the program. The major groups involved in this communication process are the staff, participants, employee committee, and management. Figure 4.1 diagrams a flowchart for communicating information to the appropriate individuals. Note that the health/fitness professional serves as a key link in the communication network. If there is not a health/fitness professional on staff, the program coordinator or chair of the employee committee can fill this role.

Staff

As Figure 4.1 depicts, the health/fitness professional receives information from and distributes results to various groups. As the intermediary, the health/fitness professional serves an important role as the

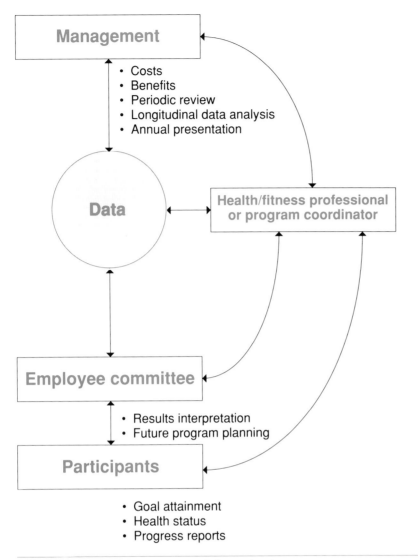

Figure 4.1 Communication network for evaluation.

interpreter of the data. She or he should manage the process of data analysis, reviewing results, comparing them to industry norms, and drawing pertinent conclusions.

These findings will catalyze brainstorming sessions that result in new programming ideas and the revision of old programs. It is important that the individuals involved in the administration of a program take part in the evaluation process. This promotes ownership of the evaluations, so they will be more effectively used to plan program expansion, develop management support, refine the program operations, and enhance staff performance.

Participants

Participants' success is primarily determined by the feedback received from staff, peers, and self. Staff members help participants set appropriate goals and measure their success. Failure to provide regular feedback often leaves participants frustrated and confused because they do not understand the changes that are occurring. This is a major

cause for dropout and low adherence rates. It is important that participant feedback be organized and provide information that is not only helpful but of high interest to the employees. Many companies use computer systems to track employee progress, then report this information on a monthly basis. Health risk assessments and annual retesting are also good feedback tools. Many times the feedback process pushes an individual into better self-discipline. Employees are also interested in how the entire employee population is benefiting from the wellness program. A summary of the annual results can be published in a newsletter to communicate the group's progress.

Employee Committee

The results should also be reviewed and discussed with the employee committee during regular meetings (e.g., quarterly or semiannually). The committee can assist the staff in interpreting findings from the employee's perspective. The employee committee members will have firsthand feedback from the other employees and can give the staff insights into marketing and programming failures, as well as help brainstorm solutions to the problems identified in the evaluation. Because evaluation serves as a springboard for future program planning, the committee needs to know and understand the results of program evaluations to assist in program development. Asking for their input also keeps them close to the program and fosters their sense of ownership.

Management

Health promotion programs communicate evaluative data to the management as a vehicle to monitor goal achievement. It is imperative that the health/fitness staff initiate this communication process and define the evaluative areas to be reported. Most senior managers have little knowledge about the workings of a health promotion program; thus it is the responsibility of the program staff to ensure that meaningful data is collected and reported. This process, over time, will educate management, fostering an understanding of and support for the program.

As discussed extensively in Phase I (p. 2), building management support is critical to long-term success. Therefore, these reports should be scheduled on a regular basis (e.g., monthly, quarterly, semi-annually, or annually). On an annual basis, a presentation should accompany the report so that the health/fitness professional and management can interact in defining new directions and strategies. The presentation should be upbeat, focusing on critical findings that are highlighted graphically with audiovisual aids. These meetings can be critical for staff to better understand management's expectations and impressions and, in turn, for management to be inspired by program growth.

Data-Base Management

The large amount of data that must be managed to successfully operate a program dictates the need for a computer system. There are many computer software programs designed to help manage and interpret health promotion program data. Even in small programs, the need for a data management system is critical to the success of the program.

Summary

Phase IV, Evaluation, relies on various techniques to review program outcomes and measure success. The approach to evaluation will depend to a great extent on the available resources, corporate objectives for the health promotion program, and the scope of the program design. Different forms of evaluation include project evaluation, quality assurance, periodic reviews, and longitudinal data analysis; each has appropriate applications to worksite health promotion. Project evaluation methods can be used to examine the success of a single program or a group of program offerings. Quality assurance, as a form of evaluation, has emerged as a well-defined process for establishing quality standards for a program and conducting routine activities to monitor whether these standards are maintained. Staff should conduct periodic reviews, on a monthly, a quarterly, a semiannual, and an annual basis, to evaluate utilization, penetration, and adherence. Longitudinal data bases will be necessary to document long-term program results, such as behavior change and cost benefits. Evaluation is meaningless unless the results are communicated to the individuals and groups involved in the program, including the staff, the participants, the employee committee, and the management.

APPENDIX A

Glossary of Practical Terms

aerobic exercise—Exercise that stimulates and conditions the cardiorespiratory system by using large muscle groups in a rhythmical, continuous manner for a minimum of 20 minutes.

aerobic fitness program—A fitness program based on aerobic exercise, designed to develop the cardiorespiratory system.

adherence—The degree to which a participant follows a prescribed program, whether attending classes or exercise sessions or conducting the program at home or work; level of adherence is an indicator of a program's success.

annual review—The final periodic evaluation in a programming year, using key results from the monthly, quarterly, and semiannual reports.

behavior change support system—Characteristics in a work environment that encourage employees to adopt healthy lifestyles; can include policies, physical environment, and corporate culture; an intervention strategy that targets the company itself, rather than the individual employees; in this book, such programs are the fourth level of intervention.

behavior modification—The alteration of behavior by arranging precedent and antecedent events to obtain a desired change in lifestyle.

behavior prescription—Lifestyle recommendations that address an individual's specific behavioral barriers to practicing healthy habits and incorporate behavior modification principles into plan for change.

breadth—The comprehensiveness of a program's design. In this book, breadth is a key to defining the program mix—it refers to the number of wellness dimensions in the overall program (e.g., preventive health, nutrition, smoking cessation, stress management, fitness).

budget—An itemized summary of proposed expenditures and income for a given period.

capital budget—An itemized summary of proposed expenditures for all items of a permanent or semipermanent nature (e.g., land, buildings, equipment).

CDC—Centers for Disease Control.

champion—One who introduces or sponsors and then advances and defends a new health/fitness program or program idea within an organization.

communication and awareness programs—Intervention strategies that disseminate health-related information using media and merchandizing techniques to influence positive health changes in a target population; in this book such programs are the first level of intervention.

compliance—The extent to which a person follows medical or health advice.

conceptual definition phase—The process of making specific decisions about a program's philosophy, scope, and design, including program mix, marketing, staffing, facility equipment, and financing; a key stage in starting a new program; in this book it is the second of four phases of development.

corporate culture—The shared goals and values of an organization, usually defined in terms of the expected, accepted, and supported norms of the group.

cost benefit analysis (CBA)—A comparison of monetary costs of programs against expected benefits; expressed in monetary units and usually stated as a ratio of benefits to costs; a positive cost benefit ratio would be greater than $1.

cost effectiveness analysis (CEA)—An evaluation of the relative costs of alternative programs for achieving the same goal; it determines how effectively and efficiently a program meets its goals, and whether altering methodology would result in greater goal realization (more effectiveness), less cost (more efficiency), neither, or both.

demographic profile—Quantifiable information on the characteristics of the employee population (e.g., age, gender, ethnicity, education level, job classification, income).

depth—The number of different program options offered in each wellness dimension; in this book, depth is a key to defining the program mix—it refers to how many of the four levels of intervention (communication and awareness, screening and assessment, education and lifestyle, and behavior change support system) are employed in a program's design.

diffusion—The process by which a program or idea is communicated through an organization over time.

education and lifestyle programs—An intervention strategy designed to give information and guidance on how to make behavior changes; in this book such programs are the third level of intervention because they provide in-depth assistance for individuals and groups on specific problems that have been identified.

emotional well-being—State in which a person is well-balanced psychologically, free of depression and anxiety, and satisfied with life.

employee advisory committee—A group of employees who assist with the development and implementation of a corporate health promotion program by representing their work groups and serving as program ambassadors.

evaluation phase—The establishment of a system of measurement procedures to provide ongoing feedback to enhance a health promotion; in this book it is the final of four phases of development.

exercise science—The technical study of exercise in health and disease; includes disciplines of kinesiology, biomechanics, exercise physiology, exercise testing, and exercise prescription.

external resource appraisal—Assessment of the degree to which consultants and packaged materials might be used in a program's design and implementation.

extrinsic reward—Positive reinforcement obtained externally (e.g., verbal recognition, prizes, and certificates) when a participant performs a desired behavior or achieves a goal.

facility plan—A detailed forecast of the space requirements, equipment needs (including a schematic), and costs of construction or renovation of space to house and equip a health promotion program.

financial plan—The component of a program proposal with a detailed budget of anticipated start-up and operating costs, including capital expenditures.

fitness assessments—Physiological tests that measure an individual's level of physical conditioning; they include cardiorespiratory tests, strength tests, flexibility tests, and body composition analysis; these tests provide baseline data to document changes resulting from an exercise program.

fitness professional—Someone with a bachelor's degree (or higher credentials) in an exercise-related field who is competent in offering exercise classes, fitness assessments, exercise prescriptions, and seminars on exercise topics; also referred to as *fitness counselor, exercise leader,* or *exercise specialist.*

goal setting—A behavior modification technique in which a participant plans how to make a lifestyle change (e.g., to stop smoking) by breaking it into a series of steps, monitoring progress, and rewarding goal achievement; typically both short-term and long-term goals are set.

health education programs—Classes, printed materials, self-help resources, bulletin boards, flyers, and so on, that provide health information on a broad range of topics, such as nutrition, fitness, prenatal care, first aid, CPR, self-care, and AIDS.

health educator—Someone with at least a bachelor's degree in health education, nutrition, nursing, or other allied health field who is trained to plan, coordinate, deliver, and evaluate health education programs.

health/fitness professional—Someone with at least a bachelor's degree (and desirably a master's) in exercise science, health education, nutrition, nursing, recreation, or allied health who plans a career devoted to promoting health. Because the field is relatively new and still evolving, there is no defined career path or entry point.

health/fitness program—An umbrella term referring to a combination of health promotion and exercise-related activities that facilitate positive lifestyle changes among participants.

health promotion—The strategies and program activities that help people change their lifestyles in the pursuit of health and well-being; includes education, behavior change, and social support strategies.

health risk profile—An assessment tool that estimates an individual's risk of becoming ill or developing certain diseases based on current lifestyle, health habits, and family history.

health screening—An assessment or combination of assessments that provides an evaluation of one's health status and detects signs or symptoms of disease.

impact evaluation—Assessment of the behavorial, attitudinal, and corporate-culture changes that occur as the result of an employee health promotion program.

implementation phase—The activation of plans to launch a health promotion program; in this book it is the third of four phases of development.

internal resource appraisal—A review of an organization's existing facilities, equipment, and staff that could be available for use in a health promotion program.

incentive system—A series of intrinsic and extrinsic rewards that are staged to increase program utilization and adherence.

initial planning phase—Collecting data and laying the groundwork for an employee health promotion program; in this book, it is the first of four phases of development.

internship—An opportunity for a student to gain practical experience and skills required of a health/fitness professional by working in a worksite health promotion setting.

intrinsic reward—Positive reinforcement obtained internally (e.g., feelings of satisfaction, self-worth, and fulfillment) when a participant performs a desired behavior or achieves a goal.

lifestyle change course—A program designed to change participant behavior; usually 8 to 12 weeks long, with meetings once or twice a week for 45 to 60 minutes.

maintenance plan—Procedures for cleaning and upkeep of facilities and equipment.

marketing—The strategies, organization, and systems that make the eligible population aware of the health and fitness program offerings, motivate them to try the program, and sustain adherence.

market analysis—Initial and ongoing assessment of the needs and interests of the target audiences, as divided into market segments.

marketing plan—An outline of the activities to be used to reach specific segments of the target audience; includes the analysis of each market segment, the development of an image and printed materials, and use of communication systems; also referred to as a *marketing strategy*.

mission statement—A testimony of the management's vision for the program, including the scope of the program and its purpose, goals, quality standards, and eligible audiences.

monthly review—The summary of a given month's daily and weekly statistics on program utilization, penetration, and adherence.

needs analysis—Systematic evaluation of data collected from surveys, and medical and personnel records to determine a company's opportunities to reduce costs or improve productivity by improving one or more health habits of the employee group.

operating budget—An itemized summary of anticipated revenue and expenses associated with the day-to-day operation of a health promotion program.

outcome evaluation—An assessment that measures whether a health promotion program caused the results that occurred.

passive smoking—Exposure to the smoke from another person's cigarette.

penetration—A measure of how deeply a health promotion program has penetrated into the total employee population; can be derived by dividing utilization by the total population.

periodic reviews—Quality assurance and monthly, quarterly, semiannual, and annual reviews; a core component of an evaluation strategy.

policy and procedure manual—A handbook that describes program operations step-by-step; often used to train and orient new staff.

process evaluation—An assessment of the qualitative aspects of program delivery to determine the effectiveness of a methodology.

program calendar—A schedule of the plans for various activities and special events to be offered in a programming cycle (e.g., quarter, season, year).

program design—A stage in the conceptual definition phase when the program mix, marketing strategy, staffing model, facility model, equipment needs, and financial plans are outlined, proposed, and approved.

program director—A health/fitness professional who is responsible for overseeing all aspects of program development, marketing, implementation, and evaluation.

program effectiveness—The efficacy with which a program has met its goals.

program goals—A set of statements about the purpose of an employee health promotion program that describe the expected achievements associated with the initiative. Each goal should include measurable objectives.

program mix—The total of opportunities offered to employees in a health promotion program, as defined by the depth and breadth of offerings.

program philosophy and scope—A stage in the conceptual definition phase when the management's beliefs, commitment, and expectations related to employee health promotion are synthesized to establish program priorities.

project evaluation—The assessment of specific programmatic segments, including process, outcome, and impact evaluation studies.

promotional campaign—A series of communications, events, or incentives used to promote awareness of program offerings or to increase participation.

quality assurance—The ongoing process of monitoring and measuring processes and outcomes against standards.

quality assurance checks—A system of measurements made during program delivery to monitor interim and final outcomes so as to maintain uniform quality.

quarterly (and semiannual) reviews—Periodic reports that allow analysis of the interaction and integration of several program activities; can be used as markers toward accomplishment of annual goals and objectives.

risk classification—A screening process that identifies one's existing and potential health problems; because of the risks and liabilities associated with exercise, it is a critical step in admitting participants to an employee fitness program.

screening and assessment programs—An intervention strategy designed to identify past, current, and potential health problems and to evaluate individuals' health and fitness levels; in this book, such programs are the second level of intervention.

self-responsibility—Assuming accountability for one's own actions and behavior change.

smoking cessation program—An intervention to help smokers quit or reduce their smoking habits.

social support system—A strategy that uses family members, co-workers, program staff, and friends to influence positively a participant's coping skills during behavior change.

staffing model—A part of the program design plan that defines the qualifications and number of staff required to manage and deliver a given program mix effectively to the eligible population.

stress management program—An intervention designed to help individuals manage stress more effectively; may address such topics as relaxation, coping skills, time management, social support, nutrition, and exercise.

support staff—Those who assist a program or facility, typically in administration and facility upkeep.

target market—The audience to whom a company aims a specific promotional and program strategy.

utilization—The number of participants in program activities for a given period.

wellness—The process of adapting patterns of behavior to lead toward health, heightened life satisfaction, and the optimal integration of social, mental, emotional, spiritual, and physical health.

Guidelines for Management Survey

Outline for a Manager Interview Process

I. Philosophy and attitudes toward health and fitness
    A. Do the managers believe that an employee health promotion program results in any of the following benefits?
        1. Improved morale
        2. Improved productivity
        3. Reduced absenteeism
        4. Reduced health care costs
        5. Recruiting advantage
        6. Improved company image
        7. Other (specify)
    B. Do the managers support the concept of offering a health promotion for the employees?*
        1. If so, to what degree?
            a. Will they participate?
            b. Will they encourage their subordinates to participate?
            c. Will they assist with program promotions and play a visible role in marketing the program?
            d. Will they serve a term on the Employee Advisory Committee?
        2. If not, why are they opposed? Probe to try to identify the source of resistance.
II. Goals and objectives for an employee health promotion program
    A. What expectations do the managers have for the health promotion program?
        1. Refer back to the potential benefits identified in I.A.
        2. In what time frame do they expect to see these outcomes demonstrated? To what degree?
        3. How realistic are their expectations, considering the resources available for intervention?

*Some of these questions have been derived from *Design of Workplace Health Promotion Programs* (p. 28) by M.P. O'Donnell, 1986, Royal Oak, MI: American Journal of Health Promotion.

B. What specific goals and objectives does the management group want to see accomplished by the program? (It might be useful to have them react to a draft of the program goals; some potential goals are identified here.)
  1. Smoking control
     a. Increase awareness about the risks of smoking.
     b. Reduce the percentage of employees who smoke.
     c. Eliminate on-the-job smoking.
     d. Become a smoke-free worksite.
     e. Hire only nonsmokers.
  2. Nutrition/weight control
     a. Reduce the percentage of overweight employees.
     b. Decrease the blood cholesterol levels of employees who have elevated cholesterol.
     c. Improve the eating habits of employees.
     d. Enhance the nutritional quality of food available at the worksite.
  3. Exercise/fitness
     a. Improve employee fitness levels.
     b. Increase the number of employees who exercise regularly.
     c. Reduce the incidence of back injuries.
     d. Increase employees' low back flexibility and strength.
     e. Achieve a certain participation level in fitness events.
     f. Enroll a certain percentage of employees in the fitness center.
  4. Stress management
     a. Decrease the level of tension and anxiety reported by employees.
     b. Improve job satisfaction.
     c. Improve employees' time management skills.
     d. Improve employees' coping skills.
  5. Other goals (specify)
III. Personal health/fitness habits
  A. Fitness habits
     1. What percentage of the managers currently exercise regularly?
     2. How often do they exercise?
     3. What types of exercise do they prefer?
  B. Nutrition/weight control
     1. What percentage of the managers would like to decrease or increase their weight?
     2. How important is it to the managers to have healthy food choices available?
     3. Do they currently make a conscious effort to select healthy foods?
     4. Are they interested in learning more about the role of nutrition in health?
  C. Smoking
     1. What percentage of the managers smoke?
     2. Of those who smoke, how much? Would they like to quit?
     3. How many are exsmokers? How did they quit?

D. Stress management
   1. What percentage of the managers feel they are under a lot of stress?
   2. What do they do to cope with stress?
   3. Do they practice any relaxation techniques regularly?
   4. What forms of recreation do they enjoy outside of work?
   5. Would they like to learn more about stress reduction techniques?
IV. Special concerns in offering worksite programs
   A. Program logistics
      1. Are there any scheduling issues to be accommodated (e.g., shifts, production lines)?
      2. Are there any space or facility limitations (e.g., conference room availability)?
      3. How does each manager want information disseminated to his or her employees?
      4. What ideas do they have for marketing the program?
   B. What barriers do the managers see to the program's success?
   C. What recommendations do the managers have to enhance the program's success?

Sample
Management Questionnaire

We are currently undertaking a study to determine the amount of interest and the kinds of feelings and assumptions that employees have about the development of a worksite health promotion program. Please answer the questions honestly. The survey is completely confidential. You do not need to give us your name.

	Agree	Disagree
It is cheaper to prevent disease than to treat it after the fact.	_____	_____
People need accurate information and education about		
a. their health risks	_____	_____
b. behaviors that create health risks	_____	_____
c. health care costs	_____	_____
d. health choices	_____	_____
e. how to change their health-related behaviors	_____	_____
People will choose to change their behavior if they are informed, motivated, and supported.	_____	_____
Healthy people do their best and are more productive on and off the job.	_____	_____
The people I associate with have an influence on my choices.	_____	_____
My work environment has an impact on my health, my behaviors, and my choices.	_____	_____

Because of the influence of the company's work environment, I have changed or I have seen co-workers change the following:

1. Start and maintain a regular exercise program	T	F
2. Stop or cut back on smoking	T	F
3. Develop skills to manage stress	T	F
4. Adopt new eating habits to maintain healthy body weight	T	F
5. Adopt new eating habits to lower cholesterol	T	F
6. Avoid the overuse of caffeine, sugar, or salt	T	F
7. Avoid the overuse or misuse of alcohol and drugs	T	F
8. Have regular medical and dental check-ups	T	F
9. Maintain healthy blood pressure	T	F
10. Understand the importance of and need for good mental and emotional health as well as physical health	T	F

What is your reaction to the prospect of a worksite health promotion program in our company?

- ☐ Excited
- ☐ Moderately interested
- ☐ Neutral
- ☐ Slightly disinterested
- ☐ Opposed

If a worksite health promotion program is implemented here, would you

a. personally participate in any programs or activities?   Y   N

b. encourage the employees you supervise to participate?   Y   N

Note. Developed by and used with permission from the Wellness Council of the Midlands (WELCOM) and reprinted from *Healthy, Wealthy and Wise: A How-to Guide for Worksite Health Promotion Managers,* a publication of the Wellness Councils of America.

Sample Employee Surveys

Employee Health Promotion Survey

We are considering the development of an employee health promotion program and would like to learn more about your interests in health and fitness. [Elaborate on the program concept in hypothetical terms, so the employees get a view of what they might expect.] Your responses will be used in planning the program and deciding what types of activities should be included.

Please take a few minutes to complete this survey. It is not necessary for you to put your name on the survey, since individual responses will be confidential.

From the following list of programs and activities, circle the number that shows your level of interest for each, with *1* being the lowest level and *5* the highest.

I. PROGRAMS

Priority

Lowest				*Highest*	A. Understanding personal health
1	2	3	4	5	1. Nutrition
1	2	3	4	5	2. Healthy lifestyle
1	2	3	4	5	3. Physical fitness education
1	2	3	4	5	4. Alcohol and other drug control
1	2	3	4	5	5. Healthy back
1	2	3	4	5	6. Men's health issues
1	2	3	4	5	7. Women's health issues
1	2	3	4	5	8. Stress management
1	2	3	4	5	9. Blood pressure management

Lowest				*Highest*	B. Reducing risks
1	2	3	4	5	1. Safety and accident prevention
1	2	3	4	5	a. Home
1	2	3	4	5	b. Gun
1	2	3	4	5	c. Water
1	2	3	4	5	d. Automobile
1	2	3	4	5	e. Motorcycle
1	2	3	4	5	f. Other _____

Priority

Lowest   Highest

1	2	3	4	5	2. Cancer risk reduction
1	2	3	4	5	3. Dental disease prevention
1	2	3	4	5	4. Heart attack risk reduction

Lowest   Highest   C. Developing healthy relations with others

1	2	3	4	5	1. Caring for and understanding aging parents
1	2	3	4	5	2. Parenting issues: Caring for and understanding children
1	2	3	4	5	3. Dealing with difficult people
1	2	3	4	5	4. Positive mental attitude

II. ACTIVITIES

Priority

Lowest   Highest   A. Promoting health through actions

1	2	3	4	5	1. Physical fitness activities (Circle any physical fitness activity you would like to take part in.)
1	2	3	4	5	a. Aerobic (exercises that bring the heart rate up to a certain level for a period of time)
1	2	3	4	5	b. Calisthenics (exercises that increase strength, balance, coordination, and joint movement)
1	2	3	4	5	c. Flexibility and stretching (exercises that increase blood supply to the muscles and improve range of motion)
1	2	3	4	5	d. Walking/jogging
1	2	3	4	5	e. Other _____
1	2	3	4	5	2. Smoking cessation
1	2	3	4	5	3. Weight management
1	2	3	4	5	4. Arthritis (help for self and family)
1	2	3	4	5	5. Blood pressure control (managing high blood pressure)

Lowest   Highest   B. Screening for specific health concerns

1	2	3	4	5	1. Glaucoma
1	2	3	4	5	2. Cholesterol
1	2	3	4	5	3. Blood pressure
1	2	3	4	5	4. Cancer
1	2	3	4	5	5. Back problems

Lowest   Highest   C. Developing skills to help others

1	2	3	4	5	1. CPR (cardiopulmonary resuscitation)
1	2	3	4	5	2. First aid

Would you attend one or more of these programs if they were offered at a convenient time?

☐ YES   ☐ NO

III. ADDITIONAL CONSIDERATIONS

Would you prefer a health promotion program at the worksite or some other place? (If other, please write down the location you would prefer.)

☐ WORKSITE
☐ OTHER (WHERE?)
_____
_____

Would your spouse or family take part in a health promotion program?

☐ YES   ☐ NO

Would you be willing to share in the cost for some programs?

☐ YES   ☐ NO

Would you take part in a weekend program?   ☐ YES   ☐ NO
Would you take part in a lunch hour program?   ☐ YES   ☐ NO

What hours do you work?   _____ a.m./p.m. to _____ a.m./p.m.

What hours are best for you to take part in a health promotion program?   _____ a.m./p.m. to _____ a.m./p.m.

In the space below, let us know about any other health care or health promotion ideas or concerns that you may have.

_____
_____
_____
_____
_____
_____
_____
_____

Return survey to _____ by _____

Thank you!

Note. Used with permission from the Wellness Council of Southeastern Wisconsin and reprinted from *Healthy, Wealthy and Wise: A How-to Guide for Worksite Health Promotion Managers*, a publication of the Wellness Councils of America.

Employee Wellness Survey

Please take a few moments to fill this out.

1. From the following list of activities and programs, check the ones you would be most interested in and if you would attend them.

Interested	Will Attend	
_____	_____	1. Alcohol and drug abuse control
_____	_____	2. Assertiveness training
_____	_____	3. Cancer risk reduction
_____	_____	4. Cardiopulmonary resuscitation (CPR)
_____	_____	5. Positive thinking
_____	_____	6. Dental disease prevention
_____	_____	7. First aid
_____	_____	8. Health awareness program
_____	_____	9. Glaucoma screening
_____	_____	10. Goal achievement
_____	_____	11. Healthy back
_____	_____	12. Heart attack risk reduction
_____	_____	13. Hypertension control
_____	_____	14. Nutrition
_____	_____	15. Physical fitness activities
_____	_____	16. Safety and accident prevention
_____	_____	17. Stress management
_____	_____	18. Weight management
_____	_____	19. Physical fitness
_____	_____	20. Other _____
		_____
		_____

2. Would you attend one or more of these programs if they were offered at a convenient time? _____ Yes _____ No _____ Maybe
   What time is most convenient for you? _____

3. Do you have a planned, regular program of exercise (swimming, walking, jogging, exercise machines) in which you participate at least three times a week? _____ Yes _____ No

4. Would you like to participate in an exercise/fitness program that was geared to your level of fitness? _____ Yes _____ No

5. What activities would you like to learn?

   _____
   _____
   _____

6. What types of exercise programs would you like to see at the worksite? _____ Aerobic _____ Jazzercize _____ Calisthenics _____ Other _____

7. Would your spouse or family participate in a health promotion program at your worksite if invited? _____ Yes _____ No

8. If one or more of the programs or activities that you selected as the most interesting to attend were offered at a convenient time and at a reasonable cost, would you probably attend that program? _____ Yes _____ No

9. Would you feel comfortable participating in a program with your co-workers? _____ Yes _____ No

10. In the space below, write any other health care or health promotion ideas or concerns that may have been triggered by the previous questions.

_____
_____
_____
_____
_____
_____
_____
_____
_____
_____
_____
_____
_____
_____

What hours do you work? _____ a.m./p.m. to _____ a.m./p.m.

Are you _____ male or _____ female?

What age group are you in?

_____ Over 20 _____ 31-40 _____ 41-50 _____ Over 50

Are you _____ administrative or _____ shop?

Note. Adapted and used with permission from Valmont Industries, Inc., Valley, Nebraska, and reprinted from *Healthy, Wealthy and Wise: A How-to Guide for Worksite Health Promotion Managers,* a publication of the Wellness Councils of America.

APPENDIX D

Resource List

Certification Programs

American Council on Exercise (formerly the IDEA Foundation)
6190 Cornerstone Court East, Suite 202
San Diego, CA 92121-4729
(800) 825-3636
Certified Aerobics Instructor
Certified Personal Trainer

Aerobics & Fitness Association of America (AFAA)
15250 Ventura Boulevard, Suite 310
Sherman Oaks, CA 91403
(818) 905-0040
Certified Aerobics Instructor

YMCA of the USA
101 North Wacker Drive
Chicago, IL 60606
(312) 977-0031
More than 90 certification programs for health/fitness professionals

Commission on Dietetic Registration
The American Dietetic Association (ADA)
216 West Jackson Boulevard, Suite 800
Chicago, IL 60606-6995
(312) 899-0040
Registered Dietitian

American College of Sports Medicine (ACSM)
P.O. Box 1440
Indianapolis, IN 46206-1440
(317) 637-9200
Health/Fitness Track
  Exercise Leader
  Health/Fitness Instructor
  Health/Fitness Director

Rehabilitative Track
  Exercise Test Technologist
  Exercise Specialist
  Exercise Program Director

National Commission for Health Education Credentialing, Inc.
Professional Examination Service
475 Riverside Drive
New York, NY 10115
(212) 870-2047
Certified Health Education Specialist

Career Placement/Recruiting Sources

Association for Worksite Health Promotion
60 Revere Drive, Suite 500
Northbrook, IL 60062
(708) 480-9574

American College of Sports Medicine (ACSM)
P.O. Box 1440
Indianapolis, IN 46206-1440
(317) 637-9200
Career opportunities bulletin for members

National Wellness Institute (NWI)
South Hall, 1319 Fremont Street
Stevens Point, WI 54481
(715) 346-2172
Monthly job opportunity bulletin for members

Association for the Advancement of Health Education (AAHE)
1900 Association Drive
Reston, VA 22091
(703) 476-3437
Computerized job bank for members

Professional Associations

Rehabilitation/Sports Medicine

American College of Sports Medicine (ACSM)
P.O. Box 1440
Indianapolis, IN 46206-1440
(317) 637-9200

Nutrition

The Sports and Cardiovascular Nutritionists
  Dietetic Practice Group (SCAN)
The American Dietetic Association
216 West Jackson Boulevard, Suite 800
Chicago, IL 60606-6995
(312) 899-0040

The American Dietetic Association (ADA)
216 West Jackson Boulevard, Suite 800
Chicago, IL 60606-6995
(312) 899-0040

Corporate Health & Fitness

Association for Worksite Health Promotion
60 Revere Drive, Suite 500
Northbrook, IL 60062
(708) 480-9574

Wellness Councils of America (WELCOA)
Historic Library Plaza
1823 Harney Street, Suite 201
Omaha, NE 68102
(402) 444-1711

Community Wellness

National Wellness Institute (NWI)
South Hall, 1319 Fremont Street
Stevens Point, WI 54481
(715) 346-2172

Exercise & Fitness

IDEA, Inc., The Association for Fitness
  Professionals
6190 Cornerstone Court East, Suite 204
San Diego, CA 92121-3773
(800) 999-4332

Public Health

Society for Public Health Education (SOPHE)
2001 Addison Street, Suite 200
Berkeley, CA 94704
(415) 644-9242

School/University

American Alliance for Health, Physical
  Education, Recreation and Dance (AAHPERD)
1900 Association Drive
Reston, VA 22091
(703) 476-3403
Six organizations under AAHPERD umbrella:
  Association for the Advancement of Health
    Education
  National Dance Association
  National Association for Sport and Physical
    Education
  National Association for Girls and Women in
    Sport
  American Association for Leisure and Recreation
  Association for Research, Administration,
    Professional Council and Societies

Provider Directories

Association for Worksite Health Promotion
60 Revere Drive, Suite 500
Northbrook, IL 60062
(708) 480-9574
Annual Membership Directory and Buyer's
  Guide free to members.

Fitness Management
3923 West Sixth Street
Los Angeles, CA 90020
(213) 385-3926
*Fitness Management Product and Services Source
  Book.* Published annually; $45.

American Hospital Association
Division of Ambulatory Care and Health
  Promotion
840 North Lake Shore Drive
Chicago, IL 60611
(312) 280-6000
*Packaged Health Promotion Program Listing,
  1990.* Free to AHA members; $10 for
  nonmembers.
*Hospital/Worksite Health Services Resource,
  1990.* $20 for members and nonmembers.

Club Industry
1415 Beacon Street, C9122
Boston, MA 02146
(800) 541-7706
(617) 277-3823
January issue Buyer's Guide. Published annually;
  free to *Club Industry* magazine subscribers.

Free and Low-Cost Program Materials

American Heart Association
7320 Greenville Avenue
Dallas, TX 75231
(214) 373-6300

American Cancer Society
777 Third Avenue
New York, NY 10017
(800) ACS-2345

National Cholesterol Education Program
National Heart, Lung, and Blood Institute
National Institutes of Health
C-200
Bethesda, MD 20892
(301) 496-4000

National High Blood Pressure Education Program
U.S. Department of Health and Human Services
Public Health Services
National Institutes of Health
Building 31, Room 4A05
Bethesda, MD 20892
(301) 496-1051

National Cancer Institute
U.S. Department of Health and Human Services
Public Health Services
National Institutes of Health
Bethesda, MD 20892
(800) 4CANCER
(301) 496-6927

APPENDIX E

Case Studies

These case studies are hypothetical cases based on actual corporate wellness programs. They are included to illustrate specific examples of design implementation approaches as support to the text. They are not intended to be used as program models. Refer to pages 28 and 29 for a discussion of these case studies.

Case I

Company Profile

Type of business:	Insurance/investment company
Annual revenue:	$180 million
Company size:	160 employees
Type of work force:	White collar and clerical
Male/female ratio:	44% men, 56% women
Age range:	22 to 65 years old, mean age = 35
Health habits:	15% smoking incidence
Education level:	Associate's/bachelor's degrees
Socioeconomic level:	Middle class, family-oriented
	Majority commute from suburbs in a large metropolitan area

Philosophy/Commitment

Program mission:	• Recruit employees to work in a transitional neighborhood of a large urban community
	• Demonstrate results of employee health promotion program on own employees to use in selling a new insurance product to other companies
	• Business strategy to differentiate company from their competition
Management commitment:	• CEO is program champion
	• Program coordination is handled by an administrative assistant in human resources department
Decision-making process:	• CEO approves program initiatives

Program Design

Program mix:	• Aerobics classes • Smoking policy • Lifestyle courses (e.g., smoking cessation, weight control, stress management) • Informal noontime walking and jogging groups
Marketing/promotions:	• Memos, fliers • Word of mouth
Staffing model:	• Independent contractors for aerobics classes • Local hospital provider for lifestyle courses and consulting • Human resources staff responsible for program coordination/marketing
Facility/equipment:	• 1,500-square-foot exercise room • Weight and cardiovascular equipment

Program Analysis

Program outcomes:	• 25% participation rate for weight control • 30% smoking cessation rate • 50% reduced smoking by 50% • 45% participation in aerobics classes
Unique characteristics:	• Top-level commitment • Strong corporate buy-in • Financial resources
Critical success factors:	• Committed CEO • Leadership style of CEO • Supportive corporate culture
Weaknesses:	• Program coordination/marketing is not a priority for human resources department • Some middle/top managers are resistant to smoking cessation policy

Case II

Company Profile

Type of business:	Bank
Annual revenue:	$1 billion
Company size:	500 employees
Type of work force:	White collar and clerical
Male/female ratio:	65% men, 35% women
Age range:	22 to 65 years old, mean age = 35
Health habits:	High incidence of cardiovascular risk in upper management Health/fitness-oriented younger work force
Education level:	Bachelor's/master's degrees
Socioeconomic level:	Young, upwardly mobile Singles and young marrieds

Philosophy/Commitment

Program mission:	• Conduct pilot study on training and development department • Use pilot study results to evaluate potential for developing a company-wide program • Compete with other local banks for talented employees
Management commitment:	• Tentative • Need to be convinced of program outcomes before investing significantly • Not a priority project
Decision-making process:	• Employee wellness committee selects program providers • Final approval for program specifics is handled by VP of training and development • Budget and overall program scope is approved by top-management committee

Program Design

Program mix:	• Health risk profile • Fitness assessments • Miscellaneous lectures • Weight control courses
Marketing/promotions:	• Memos, fliers • Promotional lectures • Employee committee
Staffing model:	• Local hospital provider • Employee wellness committee • Program coordination by secretary of training and development department
Facility/equipment:	• None

Program Analysis

Program outcomes:	• 18% participation rate health risk appraisal (HRA) • 5% participation rate in fitness assessments • 5% participation rate for weight control • 1% attendance at lectures
Unique characteristics:	• Training and development department spearheading program • Employee committee making decisions and handling program coordination
Critical success factors:	• Champion is member of top management group • Active employee committee
Weaknesses:	• Lack of top management commitment/support • Decision making by committee

- Inexperienced staff making technical decisions
- Insufficient investment to produce results that will demonstrate a payoff
- After 2 years, committee members are frustrated and losing enthusiasm with low-level participation

Case III

Company Profile

Type of business:	National conglomerate
Annual revenue:	$2 billion
Company size:	19,500 employees
	1,500 at corporate headquarters
	9,000 at remote locations
	7,000 in 5 large plants, with 750 to 1,500 in each
	2,000 in 20 sales/distribution centers, with 60 to 120 in each
Type of work force:	Headquarters and sales/distribution centers

- White/pink collar

Plants
- 20% white/pink collar
- 80% blue collar

Male/female ratio:	60% men, 40% women
Age range:	22 to 65 years old, mean age = 35
Health habits:	High incidence of cardiovascular risk in upper management
	Health/fitness-oriented younger workforce
Education level:	Bachelor's/master's degrees at headquarters
	12th grade or GED status in plants
Socioeconomic level:	Headquarters

- Young, upwardly mobile
- Singles and young marrieds

Plants
- Lower-middle class
- Families

Sales/distribution centers
- Middle to upper-middle class
- Singles and young marrieds

Philosophy/Commitment

Program mission:
- Market-driven
- Increase productivity
- Health care cost containment

Management commitment:
- 80% of senior management is personally involved in wellness lifestyle
- 8-year history to build belief and support
- Plant managers are resistant to program

Decision-making process:
- Program director (full-time health/fitness professional) reports to medical director
- Final approval by VP of human resources

Program Design

Program mix: Headquarters
- 10,000-square-foot fitness center
- Comprehensive program offerings
- Medical department involved in annual health screenings
- Healthy cafeteria
- No-smoking policy for 8 years

Plants
- Newsletter
- Recreational programs (softball and volleyball teams)
- Employee committees facilitated by safety coordinator
- Awareness programs (Great American Smoke-Out)

Sales/distribution centers
- Newsletter

Marketing/promotions:
- Memos, fliers
- Promotional lectures
- Employee committee

Staffing model: Headquarters
- Five full-time health/fitness professionals
- Medical director and two nurses
- External providers used for health promotion programs
- Employee wellness committee

Plants
- No dedicated staff
- Program coordination handled through headquarter's staff

Sales/distribution centers
- No dedicated staff
- Program coordination handled through headquarter's staff

Facility/equipment: Headquarters
- 10,000-square-foot fitness center
- State-of-the-art exercise equipment
- Sophisticated computer system for record keeping and evaluation
- Meeting rooms
- Audiovisual equipment
- Healthy cafeteria

Plants
- Recreation equipment
- Health and safety bulletin boards
- Healthy choices in vending machines
- 1-800-healthline

Program Analysis

Program outcomes: Headquarters
- 35% participation rate from employees
- 30% participation rate from management

Plants

                              • 20% participation rate

                            Sales/distribution centers

                              • 10% participation rate

Unique characteristics:
- Industry forerunner
- Top management support
- Depth/breadth of program mix
- Penetration rates

Critical success factors:
- Top-management support and commitment
- Sufficient investment to demonstrate pay-offs
- Successful track record
- Extensive evaluation system

Weaknesses:
- Need to increase penetration in remote locations
- Need to build support at plant-manager level
- Need to expand program mix in remote locations
- Program-delivery challenges in remote locations

APPENDIX F

Job Descriptions

Job title:	Program Director
Supervisor:	Medical Director or VP of Human Resources
Salary:	$25,000 to $50,000*
Role:	Assume the overall management of the health promotion program

Responsibilities

Program Operations

- Manage the development, implementation, and evaluation of all fitness programs.
- Manage the development and maintenance of procedure and training manuals on all program operations and equipment/facility maintenance protocols.
- Prepare and monitor the fitness center's operating plan and budget.
- Prepare and present summarized reports to company management regarding the fitness center's operations.

Marketing

- Develop promotional literature for the fitness center.
- Coordinate and implement special fitness center events and activities.
- Manage the development of quarterly program schedules to provide creative and interesting programs and activities for employees.

Staffing

- Manage and supervise the fitness center staff.
- Schedule and conduct weekly staff meetings.
- Monitor and evaluate staff performance.

General

- Coordinate the flow of information between company management and the fitness center staff.

---

*Salaries based on 1990 figures.

- Keep current with trends and issues affecting the health and fitness industry to incorporate state-of-the-art techniques and equipment into facility operations.
- Identify research topics and projects and participate in research design, data collection, and publication of scientific studies.
- Represent the company at health and wellness events and participate in selected associations involved in corporate health and wellness (e.g., the regional chapter of the Association for Fitness in Business).

Qualifications

Education

- Master's degree in health and fitness, exercise physiology, or a related area.

Experience

- Several years of experience supervising staff and managing the day-to-day operations of a fitness center.

Skills/Certification

- Professional, corporate image.
- Strong leadership skills and the ability to communicate effectively with employees at all levels of a company.
- CPR/First Aid certification required.
- ACSM Health/Fitness Director or Exercise Program Director certification preferred.

Job title: **Assistant Program Director**
Supervisor: Program Director
Salary: $22,000 to $40,000*
Role: Manage the day-to-day operations of the fitness center

Responsibilities

Program Operations

- Assist in the development, implementation, and evaluation of all fitness programs in accordance with established protocols.
- Help develop and promote quarterly program schedules to provide creative and interesting programs and activities for employees.
- Assist in the development and instruction of group exercise classes, and conduct fitness testing.
- Conduct fitness assessments, develop individualized exercise prescriptions, and provide initial and ongoing counseling for all fitness center members.
- Conduct group and individualized fitness-related programs for employees.

Equipment/Facility Operations

- Insure that the fitness center and the equipment are clean, well maintained, and secure.
- Monitor and insure compliance with equipment maintenance protocols and training procedures.

Staffing

- Conduct orientation and initial training for new fitness center staff.
- Schedule staff coverage for all hours of facility operation (e.g., vacations, emergency coverage).
- Develop procedures and controls to assure proper management of each staff member's responsibilities.
- Monitor work activities of each staff member on a daily basis.

General/Miscellaneous

- Keep up to date with industry trends to incorporate into fitness center operations.
- Participate in research design, data collection, and publication of scientific studies.

Qualifications

Education

- Master's degree in health/fitness, exercise physiology, or related area.

---

*Salaries based on 1990 figures.

Experience

- One to 2 years of experience working in a corporate, multi-purpose, or commercial fitness facility.

Skills/Certification

- CPR certification
- ACSM Exercise Specialist or Health Fitness Instructor certification preferred.
- Experience presenting health and wellness programs for adults at the worksite.
- Excellent interpersonal skills.

Job title:	Health Educator
Supervisor:	Program Director or Assistant Program Director
Salary:	$20,000 to $40,000*
Role:	Organize and conduct health education programs

Responsibilities

Operations

- Design, implement, and evaluate health education curricula (e.g., seminars, workshops, courses, self-help programs).
- Develop, distribute, and tabulate annual employee interest and needs surveys.
- Plan a quarterly calendar of events for wellness programming.
- Assist in the evaluation and selection of program providers.
- Assist in the development of materials for wellness seminars and programs.
- Coordinate the annual employee health fair.
- Develop an evaluation system for monitoring program results.
- Assist in the preparation of management reports for the wellness programs.
- Maintain the employee wellness resource center.

Marketing

- Develop and implement program promotional strategies for employee program participation.
- Write interest articles on health topics for the employee newsletter.
- Assist in the development of printed marketing materials.

Qualifications

Education

- Bachelor's degree in health education or a related area. Master's degree preferred.

Experience

- At least 2 years of experience delivering health programs and seminars within a corporate setting.

Skills

- Strong organizational skills.
- Ability to communicate effectively with all levels of employee.
- Outgoing, dynamic personality.
- Excellent writing skills.

---

*Salaries based on 1990 figures.

Job title:	Fitness Specialist
Supervisor:	Assistant Program Director
Salary:	$18,000 to $35,000*
Role:	Assist the Program Director with the day-to-day supervision and administration of the fitness center and other health promotion activities.

Responsibilities

Fitness Assessments

- Conduct fitness assessments, develop individualized exercise prescriptions, and provide initial and ongoing counseling.
- Assume primary responsibility for the development and instruction of group exercise classes.
- Assist in the delivery of group and individualized health and wellness programs.

Facility Operations

- Ensure that the exercise equipment is safe, clean, and in proper working order.
- Participate in compiling retention and usage data on members.

General

- Assist in developing promotional literature and conducting special fitness center events.
- Keep abreast of industry trends.

Qualifications

Education

- Bachelor's degree in health, physical education, recreation, or a related field.

Skills/Certification

- Ability to instruct exercise programs.
- Good interpersonal skills.
- Teaching and presentation skills.
- CPR certification required.
- ACSM Exercise Leader certification preferred.

---

*Salaries based on 1990 figures.

Job title:	Facility Attendant
Supervisor:	Assistant Program Director
Salary:	$5 to $9 per hour*
Role:	Assist in the daily operations of the employee fitness center

Responsibilities

Facility Operations

- Perform housekeeping duties, including cleaning the locker and grooming areas, laundering and folding towels and exercise clothing.
- Maintain inventory of toiletries and cleaning supplies, and communicate supply needs to staff member responsible for ordering.
- Learn and carry out facility operations procedures including lights, HVAC, water heater, security, etc.
- Help maintain safe and clean equipment and exercise areas.

Reception Desk (if applicable)

- Greet and assist members as they enter and leave the facility.
- Distribute and collect locker keys and towels from members.
- Keep front desk organized.

Qualifications

Education

- Associate's degree or working toward a bachelor's degree in health, physical education, or recreation.
- CPR certification required.
- Good interpersonal skills.

---

*Salaries based on 1990 figures.

Job title:	Administrative Assistant
Supervisor:	Program Director
Salary:	$16,000 to $22,000*
Role:	Provide administrative and clerical support for the fitness center staff.

Responsibilities

- Develop and maintain the fitness center filing system (for participants finances, personnel, etc.).
- Receive and direct all incoming calls.
- Order and maintain office supplies and equipment.
- Assist staff with preparation of required reports.
- Take minutes at weekly staff meetings, and prepare summarized report for staff members.
- Receive and distribute fitness center mail and packages.
- Assist with front reception desk duties.

Qualifications

Education

- High school education or equivalent.

Experience

- One to 2 years of experience working in a corporate, multipurpose, or commercial fitness facility.
- Experience within the company desirable.

Skills/Talents

- Good organizational skills.
- Great communication skills.
- Outgoing personality.
- Personal commitment to health/fitness.

*Salaries based on 1990 figures.

Sample Table of Contents for Policy and Procedure Manual

Key area	Subsections
Programming	Enrollment Testing/retesting Exercise prescriptions Orientation Exercise logging Master event calendar and special promotions Incentive point systems and awards
Staff management	Scheduling Temporary staff replacements Internships Contract employees Vacation requests Media inquiries
Member/participant management	Enrollment/termination procedures Participant records Inactive member follow-up Participant payment Reference library
Facility management	Opening/closing procedures Facility maintenance/cleaning Guests Emergency procedures Lost and found Illness/injury reporting Key control Purchase of supplies
Locker maintenance	Sign-up/termination Towel control Laundry Locker/storage bin control Purchase and inventory of supplies
Equipment maintenance	Cleaning and maintenance Office purchasing

Key area	Subsections
Management reports	Utilization
	Retention/participation
	Financial
Miscellaneous	Videos
	Library
	Purchases

References

1. Dunn, H. *High Level Wellness*. Beatty, Arlington, VA, 1961, p. 5.

2. Opatz, J.P. *Primer of Health Promotion: Creating Healthy Organizational Cultures*. Oryn, Washington, DC, 1985, p. 7.

3. Association for Fitness in Business. *1990-91 Who's Who in Employee Health and Fitness Directory*. Author, Indianapolis, 1990, pp. 98-110.

4. Frank, A. Wellness committee management a flop. *Optimal Health*, January/February, 1989; 4(1):38.

5. Horton, W.L. Quality pays off. *Fitness in Business*, August, 1988, pp. 10-11.

6. O'Donnell, M.P. Design of workplace health promotion programs. *American Journal of Health Promotion*, Royal Oak, MI, 1986, pp. 27, 10-11, 18-22.

7. Brox, A. Wellness: A special report. *Club Industry*, January, 1985, p. 23.

8. Sloan, R.P., Gruman, J.C., Allegrante, J.P. *Investing in Employee Health: A Guide to Effective Health Promotion in the Workplace*. Jossey-Bass, San Francisco, 1987, pp. 103-106.

9. Lefebvre, R.C., Flora, J.A. Social marketing and public health intervention. *Health Education Quarterly*, 1988; 15:299-315.

10. Stern, M.P., Farquhar, J.W., Maccoby, N., Russell, S.H. Results of a two-year health education campaign on dietary behavior: The Stanford three community study. *Circulation*, 1976; 54(5):826-833.

11. Farquhar, J.W., Wood, P.D., Breitrose, H., Haskell, W.L., Meyer, A.J., Maccoby, N., Alexander, J.K., Brown, B.W., McAlister, A.L., Nash, J.D., Stern, M.P. Community education for cardiovascular health. *Lancet*, 1977; 1:1192-1195.

12. Maccoby, N., Farquhar, J.W., Wood, P.D., Alexander, J. Reducing the risk of cardiovascular disease: Effects of a community-based campaign on knowledge and behavior. *Journal of Community Health*, 1977; 3(2):100-114.

13. Farquhar, J.W. The community-based model of life-style intervention trials. *Journal of Epidemiology*, 1978; 108(2):103-111.

14. Flay, B. Mass media and smoking cessation: A critical review. *American Journal of Public Health*, 1987; 77:153-160.

15. Flora, J., Maibach, E., Maccoby, N. The role of mass media across four levels of health promotion intervention. *Annual Review of Public Health*. L. Breslow, J. Fielding, L. Lave (Eds.). Palo Alto, CA, 1989.

16. Freimuth, V., Hammond, S., Stein, J. Health advertising: Prevention for profit. *American Journal of Public Health*, 1988; 78:557-561.

17. Vickery, D.M., Golaszewski, T.J., Wright, E.C., Kalmer, H. The effect of self-care interventions on the use of medical services within a Medicare population. *Medical Care*, 1988; 26(6):580-588.

18. DeClemente, C.C., Prochaska, J.O. Self-change and therapy change of smoking behaviors: A comparison of processes of change in cessation and maintenance. *Addictive Behaviors*, 1982; 7:133-142.

19. Storlie, J. Techniques of behavioral intervention. *Behavioral Management of Obesity*. J. Storlie, H.A. Jordan (Eds.). Human Kinetics, Champaign, IL, 1984, pp. 19-48.

20. Glasgow, R.E., Rosen, G.M. Behavioral bibliography: A review of self-help behavior therapy manuals. *Psychological Bulletin*, 1978; 85:1-23.

21. Kanfer, F.H., Goldstein, A.P. *Helping People Change*. Pergamon, New York, 1975, pp. 195-228.

22. McLeroy, K.R., Bibeau, D., Steckler, A., Glanz, K. An ecological perspective on health promotion programs. *Health Education Quarterly*, 1988; 15:351-378.

23. Harrison, M.I. *Diagnosing Organizations: Methods, Models, and Process*. Sage, Newbury Park, CA, 1987, pp. 1-22.

24. Bellingham, R., Cohen, B., Edwards, M.R., Allen, J. *The Corporate Culture Sourcebook*. Human Resources Development Press, Amherst, MA, 1990, pp. 184-192.

25. Rogers, E.M. *Diffusion of Innovations*. Free Press, New York, 1983.

26. Gerson, R.F. *Marketing Health/Fitness Services*. Human Kinetics, Champaign, IL, 1989, p. 19.

27. Horton, W.L. Sound staffing fosters success. *Fitness in Business*, December, 1988, pp. 96-97.

28. Fitness Systems. *Salary Survey Results: Corporate Fitness Center Personnel*. Fitness Systems, Los Angeles, 1989.

29. Patton, R.W., Grantham, W.C., Gerson, R.F., Gettman, L.R. *Developing and Managing Health/Fitness Facilities*. Human Kinetics, Champaign, IL, 1989, p. 135.

30. Liebler, J.G., Levine, R.E., Dervitz, H.L. *Management Principles for Health Professionals*. Aspen, Rockville, MD, 1984, p. 222.

31. Hall, J. Which health screening techniques are cost-effective? *Diagnosis*, February, 1980, p. 65.

32. American College of Sports Medicine, *Guidelines for Exercise Testing and Prescription*. 4th ed. Lea & Febiger, Philadelphia, 1991, pp. 1-9.

33. Howley, E.T., Franks, B.D. *Health/Fitness Instructor's Handbook*. Human Kinetics, Champaign, IL, 1986.

34. Franks, B.D., Howley, E.T. *Fitness Leader's Handbook*. Human Kinetics, Champaign, IL, 1989.

35. Allsen, P. *Total Fitness for Life*. Brown, Dubuque, IA, 1984.

36. Van Gelder, N. (Ed.). *Aerobic Dance-Exercise Instructor Manual*. IDEA Foundation, San Diego, 1987.

37. Dishman, R.K. *Exercise Adherence: Its Impact on Public Health*. Human Kinetics, Champaign, IL, 1987, pp. 377-379.

38. Landgreen, M., Baun, W.B. Adhering to fitness in the corporate setting. *Corporate Commentary*, 1984; 1(1):37-44.

39. Goldfried, M.R., Merbaum, M. (Eds.). *Behavior Change Through Self-Control*. Holt, Rinehart & Winston, New York, 1973.

40. Rejeski, W.J., Kenney, E.A. *Fitness Motivation: Preventing Participant Dropout*. Life Enhancement, Champaign, IL, 1989, pp. 6-8.

41. Marlatt, G.A., Gordon, J.R. *Relapse Prevention: Maintenance Strategies in the Treatment of Addictive Behaviors*. Guilford Press, New York, 1985, pp. 216-220.

42. Glanz, K., Lewis, F.M., Rimer, B.K. *Health Behavior and Health Education: Theory, Research, and Practice*. Jossey-Bass, San Francisco, 1990, pp. 161-186.

43. Blanke, K., Stanek, K., Stacy, R. Comparison of the success of nutrition education to lower dietary cholesterol and fat with and without spouse support for individuals with elevated blood cholesterol. *Health Values*, 1990; 14(3): 33-37.

44. Doherty, W.J., Schrott, H.G., Metcalfe, L., Lasiello-Vailas, L. Effect of spouse support and health beliefs on medication adherence. *The Journal of Family Practice*, 1983; 17(5):837-841.

45. Allen, R.F., Kraft, C. *Beat the System: A Way to Create More Human Environments*. McGraw-Hill, New York, 1980, pp. 170-172.

46. Wilbur, C.S. The Johnson and Johnson program. *Preventive Medicine*, 1983; 12:672-681.

47. Sandhussen, R.L. *Marketing: Barron's Business Review Series*. Barron's, New York, 1987, p. 287.

48. Baun, W.B., Baun, M.R. A corporate health and fitness program: Motivation management by computers. *Journal of Physical Education, Recreation and Dance*, 1984; 55(4):42-45.

49. Parkinson, R. Participation: Keystone in health promotion evaluation. *Corporate Commentary*, November, 1984, pp. 30-37.

50. Allen, R.F., Allen, J., Kraft, C., Certner, B. *The Organizational Unconscious: How to Create the Corporate Culture You Want and Need*. Human Resource Institute, Morristown, NJ, 1982, p. 9.

51. Chenoweth, D.H. *Planning: Health Promotion at the Worksite*. Benchmark Press, Indianapolis, 1987, pp. 111-113.

52. Franklin, B.A. Exercise program compliance. *Behavioral Management of Obesity*. J. Storlie, H.A. Jordan (Eds.). Human Kinetics, Champaign, IL, 1984, p. 106.

53. Jordan, H.A. Motivational strategies. *Behavioral Management of Obesity*. J. Storlie, H.A. Jordan (Eds.). Human Kinetics, Champaign, IL, 1984, p. 96.

54. O'Donnell, M.P. Cost-benefit analysis is not cost-effective. *American Journal of Health Promotion*, 1988; 3(1):74-75.

55. Murphy, R.J., Ellas, W.S., Gasparotto, G., Huset, R.A. Cost-benefit analysis in worksite health promotion evaluation. *Fitness in Business*, August, 1987, pp. 9-14.

Recommended Readings

Many of the ideas presented in this paper build on general concepts presented in many texts. These texts will also serve as a valuable resource to the reader who wants additional references.

Bouchard, C., Shephard, R.J., Stephens, T., Sutton, J.R., McPherson, B.D. *Exercise, Fitness, and Health: A Consensus of Current Knowledge.* Human Kinetics, Champaign, IL, 1990.

Caltaldo, M.F., Coates, T.J. *Health and Industry: A Behavioral Medicine Perspective.* Wiley, New York, 1986.

Chenoweth, D.H. *Planning Health Promotion at the Worksite.* Benchmark Press, Indianapolis, 1987.

Dunn, H. *High Level Wellness.* Beatty, Arlington, VA, 1961.

Fielding, J.E. *Corporate Health Management.* Addison-Wesley, Reading, MA, 1984.

Fiest, J., Brannon, L. *Health Psychology: An Introduction to Behavior and Health.* Wadsworth, Belmont, CA, 1988.

Kernaghan, S.G., Giloth, B.E. *Tracking the Impact of Health Promotion on Organizations: A Key to Program Survival.* American Hospital, Chicago, 1988.

King, N.J., Remenyi, A. *Health Care: A Behavioral Approach.* Grune & Stratton, New York, 1986.

Kizer, W.M. *The Healthy Workplace: A Blueprint for Corporate Action.* Wiley, New York, 1987.

Loudon, D.L., Della Britta, A.J. *Consumer Behavior: Concepts and Applications.* McGraw-Hill, New York, 1984.

McDowell, I., Newell, C. *Measuring Health: A Guide to Rating Scales and Questionnaires.* Oxford University Press, New York, 1987.

O'Donnell, M.P., Ainsworth, T.H. *Health Promotion in the Workplace.* Wiley, New York, 1984.

Opatz, J.P. *Primer of Health Promotion: Creating Healthy Organizational Cultures.* Oryn, Washington, DC, 1985.

Parkinson, R.S. *Managing Health Promotion in the Workplace: Guidelines for Implementation and Evaluation.* Mayfield, Palo Alto, CA, 1982.

Patton, R.W., Corry, J.M., Gettman, L.R., Graf, J.S. *Implementing Health/Fitness Programs.* Human Kinetics, Champaign, IL, 1986.

Patton, R.W., Grantham, W.C., Gerson, R.F., Gettman, L.R. *Developing and Managing Health/Fitness Facilities.* Human Kinetics, Champaign, IL, 1989.

Rippe, J.M. *Fit for Success: Proven Strategies for Executive Health.* Prentice Hall Press, New York, 1989.

Ross, R.M., Jackson, A.S. *Exercise Concepts, Calculations, and Computer Applications.* Benchmark Press, Indianapolis, 1990.

Russell, M.L. *Behavioral Counseling in Medicine: Strategies for Modifying At-Risk Behavior.* Oxford University Press, New York, 1989.

Sallis, J.F., Hovell, M.F. Determinants of exercise behavior. *Exercise and Sport Sciences Reviews: American College of Sports Medicine Series.* Vol. 18. K.B. Pandolf (Ed.). Williams & Wilkins, Baltimore, 1990.

Sampson, E.E., Marthas, M. *Group Process for the Health Professions.* Wiley, New York, 1981.

Shephard, R.J. *Economic Benefits of Enhanced Fit-ness*. Human Kinetics, Champaign, IL, 1986.

Tubesing, N.L., Tubesing, D.A. *Structured Exercises in Wellness Promotion*. Vol. 1 & 2. Whole Person Press, Duluth, MN, 1983, 1984.

Veney, J.E., Kaluzny, A.D. *Evaluation and Decision Making for Health Services Programs*. Prentice Hall, Englewood Cliffs, NJ, 1984.

Wellness Councils of America. *Healthy, Wealthy and Wise: A How-to Guide for Worksite Health Promotion Managers* (2nd ed.). Wellness Councils of America, Omaha, NE, 1990.

Index

A

Adherence, program, 81
Aerobics and Fitness Association of America (AFAA), certification program of, 67
Aerobics studio, space requirements for, 37
American College of Sports Medicine (ACSM)
  certification program of, 67
  risk criteria of, 55
American Council on Exercise, certification program of, 67
Annual review, of program, 81
Annual survey in marketing analysis and monitoring, 65-66
Assistant program director, job description for, 115-116
Association for Fitness in Business (AFB), 139
  annual directory, 13, 15, 27
  Internship Clearinghouse of, 68
  membership benefits of, 139
  mission statement of, 139
  organization of, 139
  as resource in developing health promotion program, 4
AV equipment, in educational programs, 43

B

Behavior change, 82
Behavior change support system, 25, 27
Behavioristic bases, 65
Biofeedback, 32
Buddy Exercise System, 77
Buddy system
  in incentive program, 77
  in wellness programs, 61

C

Cafeteria-style benefits plan, 10
Calendar, in health promotion program, 50-54
Capital budget, 45-47
Cardiovascular/aerobic conditioning, 41
Case studies, 107-112

CEO

  in health promotion program, 3
  in marketing campaign, 63
  and organizational structure of health promotion program, 6-7
  and planning of promotional campaign, 31
Champion
  characteristics of, 3
  and organizational structure of health promotion program, 6
Cholesterol screening, 57
Communication and awareness programs, 24-26
Communication channels, establishing, 2-3
Company size, as factor in developing health promotion program, 28-30
Computerized health assessment tools, 13
Computers
  in developing personalized programs, 27
  in educational programs, 43-45
  in health promotion program, 71-72
Conference rooms, 40-41
Contract management company, and staffing, 34-35
Corporate culture, influence of, on employee behavior, 62
Corrective actions, 80
Cost-benefit analysis, and use of longitudinal data base, 82
Cost-effectiveness analysis, 77
Counseling, 58-60

D

Data-base management, and results interpretation and communication, 84
Demographic bases, 65
Demographic profile of employees, 13

E

Educational programs
  equipment needs of, 15, 41-45
  facility plan for companies developing, 39-41
  financial plan for, 45-48

Educational programs (*continued*)
   fitness equipment in, 41-43
   screening process in, 57
Education and lifestyle programs, 25, 27
Education-oriented program, staffing level in, 33
Employee committee
   in developing health promotion program, 5
   in launching marketing campaign, 64-65
   marketing role of, 63
   and results interpretation and communication, 84
Employee health promotion survey, 14, 97-101
Employee population, in needs analysis, 13-14
Enrollment and health screening, 54-57
Equipment needs, 15, 41-45
Evaluation plan, 22
External resource appraisal, 15-16

F

Facilities and equipment, in internal resource appraisal, 14-15
Facility and equipment maintenance, 68-69
Facility attendant, job description for, 119-120
Facility plan
   for educational programs, 39-41
   for fitness centers, 36-39
Financial commitment, securing preliminary, 7-10
Financial plan, 45
   capital budget in, 45-47
   operating budget in, 47-48
Fitness centers
   facility plan for companies developing, 36-39
   layout and design considerations, 39
   space requirements, 36-38
   staffing of, 33
Fitness equipment, in educational program, 41-43
Fitness/health promotion interests and habits, 14
Fitness professionals, 35
Fitness retesting standards, 57
Fitness Systems, 36
Freelance instructors, 35
Free-weights, proper use of, 42

G

Goals, 19-21, 60
Great American Smoke Out, 52, 111

H

Health assessment equipment, 45
Health educator
   job description for, 117-118
   responsibilities of, 35-36
Health/fitness professional
   criteria for selection of, 4
   in launching marketing campaign, 63-64
   qualifications of, 32
   responsibility for health promotion program, 7
   as role model in wellness programs, 61
Health promotion, vii
Health promotion program
   activation of, 50-54
   calendar for, 50-54
   conceptual definition in, ix, 17-48

   counseling and orientation in, 58-60
   enrollment and health screening in, 54-57
   evaluation of, x, 22, 73-85
   forms and protocols for, 70-71
   goals of, 19-21, 60
   health/fitness assessments in, 57-58
   implementation of, ix-x, 49-72
   incentive system in, 60-61
   initial planning in, viii-ix, 1-16
   length of, 51-52
   longitudinal data analysis in, 81-82
   management commitment and support for, 2-10
   marketing strategy for, 30-32, 62-66
   mission statement of, 19
   needs analysis for, 10-16
   operations and administration for, 68-72
   organizational structure for, 5-7
   periodic reviews in, 77-81
   phases in, viii-x
   philosophy and scope of, 18-19
   priorities in, 21-22
   program design in, 22-23
   program mix in, 23-30
   quality standards for, xi
   registration in, 54-55, 56
   results interpretation and communication in, 82-84
   retention and motivation in, 60-62
   risk classification in, 55-57
   seasonal cycles in, 52
   securing preliminary financial commitment in, 7-10
   small- versus large-company focus in, 28-30
   social support system in, 61-62
   special events in, 52
   staffing model in, 32-36
   staff selection in, 66-68
   types of, vii-viii
   work-load cycles in, 52, 54
Health risk profile of employees, 13
High Energy Incentive Program, 77
Human resources department, responsibility for health promotion program, 6-7

I

Image development, 30-31
Incentive program, 60-61, 65
Internal resource appraisal, 14-15
Internship programs, 68
   staffing for, 33-34

J

Job classifications, 35-36
Job descriptions, 113-120

K

Kickoff event, 63

L

Locker rooms, space requirements for, 38
Longitudinal data analysis, 81-82
   behavior change, 82
   cost-benefit analysis, 82

M

Management
   access to top, 7-10
   establishment of communication channels with,
     2-10
   and image development, 30
   in launching marketing campaign, 63
   and results interpretation and communication, 84
   securing financial commitment, 7-10
Management by objectives approach, 81
Management survey, guidelines for, 91-95
Manager interview process, outline for, 91-93
Managers, in needs analysis, 11-12
Market analysis and monitoring, 65-66
Marketing campaign, launching, 63-65
Marketing strategy, 30, 62-63
   defining target market in, 30
   image development in, 30-31
   launching marketing campaign in, 63-65
   market analysis and monitoring in, 65-66
   promotional campaign in, 31-32
   role of employee committee in, 64-65
   role of health/fitness professionals in, 63-64
   role of top management in, 63
Medical care cost data, 14
Medical equipment, 15
Mission statement, 19
Monthly review, of program, 80-81

N

Needs analysis
   employee population in, 13-14
   external resource appraisal, 15-16
   internal resource appraisal, 14-15
   managers in, 11-12
   purpose of, 10
   sources of data for, 11
Nutrition education programs, 15, 45

O

Operating budget, 47-48
Operations and administration, 68
   facility and equipment maintenance, 68-69
   policy and procedure manual, 68
   record keeping, 70-72
Organizational structure, 5-7
Orientation, 58-60
Outcome evaluation, 74-76

P

Participants, and results interpretation and communication, 83-84
Participation rates, 21-22
Penetration, 80-81
Periodic reviews, of program, 77-81
Personal health/fitness habits, of managers, 12
Philosophy, 18-19
Policy and procedure manual, 68
   sample table of contents for, 121-122
Preventive maintenance, 68-69
Priorities, 21-22

Process evaluation, 76
Program breadth, 23, 24
Program depth, 23, 24-28
Program design, 22-23
Program director, job description for, 113-114
Program mix, 23-30, 32-33
   as factor in facility design, 39
   preprogram assessments in, 57
Program sponsor, 2
   characteristics of, 3
   role of, 3
Project evaluation, 74
   annual review, 81
   corrective actions, 80
   cost-effectiveness analysis, 77
   impact evaluation, 74, 76
   longitudinal data analysis, 81-82
   monthly review, 80-81
   outcome evaluation, 74-76
   periodic reviews, 77-81
   process evaluation, 76-77
   quality assurance, 78-80
   quality standards, 78, 79
   quarterly reviews, 81
   results interpretation and communication, 82-84
   semiannual reviews, 81
Psychographic bases, 65

Q

Quality assurance, 78-80
Quality standards, for health promotion program, xi,
   78, 79
Quarterly reviews, of program, 81

R

"Reach and repeat" concept, 24-25
Record keeping
   computers in, 71-72
   forms and protocols in, 70-71
   in marketing analysis and monitoring, 65
Referral program, role of employee committee in, 65
Registration, 54-55, 56
Resource list, 103-105
Retention and motivation, 60-62
Risk classification, 55-57
Risk screening, 32

S

Scope, 18-19
Screening and assessment programs, 25, 26-27
Seasonal cycles, 52
Semiannual review, of program, 81
Smoking cessation, viii
Social support system, 61-62
Space requirements, for health promotion programs,
   36-38
Special events, 52, 63
   role of employee committee in, 64-65
Sponsoring organization, role of, in wellness
   process, vii
Spotters, 42

Staff
    as factor in facility design, 39
    hiring or contracting, 34-35
    in internal resource appraisal, 15
    for internship programs, 68
    job classifications, 35-36
    qualifications and certifications, 66-67
    recruitment of, 67
    and results interpretation and communication,
      82-83
    staffing level, 32-34
Strength-training equipment, 41-42
Suggestion box, 70
Support staff, 36
T
Target market analysis, 62-63
Target market definition, 30

Technical expert
    in developing health promotion program, 4-5
    in facility design, 39
U
Utilization, 80
Utilization statistics, 72
W
Waiting list, 54-55
Walking trails, 14-15
Wellness, definition of, vii
Wellness Councils of America, 104
Wellness philosophy and attitudes of managers, 12
Work-load cycles, 52, 54

About the Authors

Authors

Jean Storlie, MS, RD
President, Jean Storlie Associates

Jean Storlie is a registered dietician with extensive experience in nutrition, fitness, and health promotion. She received an MS in adult fitness/ cardiac rehabilitation from the University of Wisconsin-La Crosse, where she was the director of nutrition services for the La Crosse Exercise Program from 1979 to 1983. Since then Jean has been involved in the design and delivery of corporate wellness programs through Dr. Cooper's Aerobics Center in Dallas and Rush-Presbyterian-St. Luke's Medical Center in Chicago. In 1990, she founded Jean Storlie Associates, a consulting practice in nutrition communications, health promotion, and sport nutrition. Storlie is the past chair of the Sports and Cardiovascular Nutrition Practice Group of the American Dietetic Association and a member of both the American College of Sports Medicine and the Association for Fitness in Business. She serves on the editorial board of the *International Journal of Sport Nutrition* and the advisory board of IDEA, the Association for Fitness Professionals. She also teaches graduate courses in sport nutrition and wellness for the Department of Clinical Nutrition at Rush University.

William B. Baun, MS
Manager, Tenneco Health and Fitness

William B. Baun is the manager of the Health and Fitness Department at Tenneco—a conglomerate made up of five major operating divisions and 96,000 employees worldwide. In this position, he is responsible for ensuring the highest quality of development, implementation, and evaluation of Tenneco's health and fitness programs. Baun received his master's degree in exercise science in 1980. He has published and presented extensively on exercise compliance/adherence and the economic benefits of worksite health promotion programs. He joined the Tenneco staff as an exercise physiologist in 1981 and in 1984 was promoted to his present position. Baun has received the Association for Fitness in Business Exceptional Leadership Award and is certified as a health/fitness director through the American College of Sports Medicine. He is working on his doctoral degree in human organization systems with The Fielding Institute in Santa Barbara, CA.

William L. Horton, MBA
President, Fitness Systems

In 1975, William L. Horton founded Fitness Systems, the leading provider of corporate health and fitness program management in the nation. After receiving his master's degree with distinction from the Harvard Graduate School of Business Administration in 1966, Horton went on to serve as an associate with a prominent management consulting firm, an executive assistant to the Secretary of Health, Education and Welfare, and a staff assistant to the President of the United States. In addition, Horton has served in several capacities in the leadership of the Association for Fitness in Business. He wrote the "Management Update" column for AFB's *Fitness in Business* and is a member of the editorial board of the *American Journal of Health Promotion*. Horton also speaks to business and academic groups on the planning and analysis of the costs and benefits of employee fitness programs.

Reviewers

Kevin M. Clair, MHA, MA

Larry R. Gettman, PhD

Robert Karch, EdD

Mark Landgreen, MA

Michael P. O'Donnell, MBA, MPH

Robert W. Patton, PhD

George Pfeiffer, MSE

Daniel J. Lynch, MS, past president of AFB (1989), conceptualized and spearheaded this project. Jonathan D. Trick, MS, vice president of special projects for AFB (1989), served as project coordinator during the initial planning.

Corporate Sponsors

The Association for Fitness in Business is fortunate to have many companies and organizations involved who believe in the promotion of employee health and fitness.

In June 1990, two companies provided grants to AFB to assist in the administrative costs to develop this publication. In addition, Tenneco, Inc., provided many of the photos used in this text. Both companies' special assistance helped to make this project possible. We certainly appreciate their tremendous support and extra effort in the promotion of employee health.

Tenneco, Inc.—Houston, Texas

Townsend Engineering, Inc.—Des Moines, Iowa

Thank you to our corporate sponsors!

TWENTY YEARS OF LEADERSHIP AND GROWTH IN WORKSITE HEALTH PROMOTION

Since 1974, the Association for Worksite Health Promotion championed worksite health promotion as an appropriate and effective business strategy. In 1994, we are celebrating 20 years of helping worksite health promotion professionals develop their skills, enhance their careers and improve the effectiveness of the programs they manage.

Recent AWHP initiatives include *AWHP's Worksite Health*, specifically designed to meet the needs of the practicing worksite health professional; publications like the *Worksite Health Professionals National Compensation and Benefits Survey* and *101 Fitness Programs*; increased legislative affairs activities; the development of professional standards for worksite health practitioners; and 12 topical Practice Groups to address the specific issues in our increasingly sophisticated field.

Of course, the focal point of every year is the AWHP Annual International Conference. Upcoming conference dates are:

1995 – Stouffer Orlando Resort, Orlando, Florida;

1996 – Hilton Pointe on South Mountain, Phoenix, Arizona;

1997 – Palmer House Hilton, Chicago, Illinois

What does the next 20 years hold for worksite health promotion professionals? What new challenges lie ahead? One thing is certain — our field is changing. AWHP's members, sponsors and friends are committed to meeting these challenges. Join us!

ASSOCIATION FOR
Worksite Health
PROMOTION

FOR MORE INFORMATION ABOUT AWHP, CALL ZOIE GELEERD AT
(847) 480-9574

ASSOCIATION FOR
Worksite Health
P R O M O T I O N

FACT SHEET

Mission: To advance worksite health promotion throughout the world. The Association for Worksite Health Promotion (AWHP) is dedicated to enhancing the personal and organizational health and well-being of employees and their families. We work to achieve our mission by: advocating the value of worksite health promotion to business and government leaders; supporting health promotion professionals through education; providing resources to those who offer health promotion at the worksite; and serving as a catalyst to advance research and learning in our field.

History: In 1974, the organization was founded as the American Association of Fitness Directors in Business and Industry and in 1983 the name was changed to the Association for Fitness in Business. In 1993, the organization was named the Association for Worksite Health Promotion (AWHP) which better represents the Association's evolution and its comprehensive approach to worksite health promotion.

Structure: AWHP is an international association with eight regional U.S. chapters (each with an elected Board of Directors), two Canadian chapters and a chapter in the United Kingdom. Regional chapters hold activities such as conferences, state meetings and workshops at reduced rates to AWHP members. AWHP also has international members in Australia, Brazil, Germany, Indonesia, Israel, Japan, Malaysia, the Netherlands, New Zealand, South Africa, Sweden, Taiwan, Trinidad and Tobago and the United Kingdom.

Membership: As of December 31, 1995, AWHP has approximately 2,600 members divided into four membership classes: professional, company, associate and student. Members include human resource directors, health educators, corporate and nonprofit organization wellness directors, government officials, exercise physiologists, dieticians, organizational health consultants, benefits managers, physical and occupational therapists, occupational health nurses and physicians, fitness instructors, personal trainers, and exercise facility managers/owners.

Benefits: Membership in AWHP is recognized as the standard of professionalism within the worksite health promotion community. AWHP members get the latest information on the worksite health promotion industry from the AWHP *ACTION* newsletter, a members-only publication, and *AWHP's Worksite Health*, the first magazine focusing on practical information written for practicing worksite health promotion professionals. In addition, members receive reduced registration fees to attend the AWHP International Conference, the largest event in the world dedicated exclusively to worksite health promotion. In 1994, AWHP introduced "Practice Groups," an additional networking resource allowing members to meet with other AWHP professionals in the same field. PRINT RESOURCES: AWHP members receive discounts on publications such as the *1993 National Compensation and Benefits Survey, Economic Impact of Worksite Health Promotion* and *101 Fitness Programs for Organizational Health*. In addition, only members receive the AWHP Membership Directory: a listing of AWHP members, worksite health suppliers, university programs, local chapter information and more.

Location: AWHP's main office is located at 60 Revere Drive, Suite 500, Northbrook, IL 60062. For more information, or to receive a membership brochure or other publications, please call our office at (847) 480-9574.

ASSOCIATION FOR
Worksite Health
PROMOTION

MEMBERSHIP FACT SHEET

AWHP Membership Benefits

Individual members and the official representatives for the company memberships will receive these "tangible" benefits:

■ A subscription to *AWHP's Worksite Health*;
■ A subscription to the AWHP *ACTION* newsletter;
■ The annual Membership Directory and Buyer's Guide (an exclusive benefit) and free inclusion if membership status is current as of Feb. 1;
■ Automatic membership into regional and local sections;
■ Reduced subscription rates for the JOB Opportunity Bureau; and
■ Preferred rates for AWHP International and Regional Conferences and Events.

In addition, the Association will offer and arrange special programs or services for AWHP members only.

Associate member companies will receive the following additional benefits:

■ Eligible to exhibit at the Annual Conference; and
■ Preferred rate (an additional 10% reduction of the low member price) on mailing list rental.

AWHP Membership Categories

The association has a category of membership available to meet your situation. Following is a listing of the types of memberships provided. Please note, only the Professional Member has the right to hold office or vote on Association matters.

1. **Individual Memberships**—these memberships are in the name of the individual rather than the company or organization.

   A) **Professional Member**—is for individuals who derive income from a health promotion profession by providing educational development, management services or evaluations of health promotion programs.

   B) **Student Member**—is for full-time undergraduate and graduate students enrolled in a program of study related to health promotion. (AWHP also provides a student chapter program at qualified institutions. Please call the AWHP office for more information.)

2. **Company Memberships**—these memberships are in the name of the firm, business or organization. The company/organization shall designate the person or persons who will represent them with the Association and shall pay annual dues for each such person. Memberships may be transferred by the company/organization upon written notice to AWHP.

   A) **Associate Member**—is available to any firm, business or corporation engaged in selling products or services to members of the Association.

   B) **Company/Organization Member**—is available to any firm, business, not-for-profit organization or institution with an active interest in health promotion or which may be helpful in carrying out the objectives of the Association.

AWHP Application for Membership
(Please type, print or attach a business card)

Date of Application _____ Name _____
                                              (First)   (M.I.)   (Last)

Title _____ Company/Organization _____

Address _____ City _____ State _____ Zip _____

Phone (_____) _____ Fax (_____) _____

> Please check the membership you are applying for and submit the appropriate annual fee in U.S. Dollars or the equivalent. (Memberships are based on the calendar year, Jan. 1-Dec. 31.)

❏ Professional Member—$130   ❏ Associate Member—$350

❏ Student Member*—$70   ❏ Company/Organization Member—$250

   *A student application must be accompanied by a letter from the registrar's office or a current transcript.

❏ Check enclosed for $ _____ ❏ Charge $_____ to my: _____ ❏ Visa ❏ MasterCard

Acct. # _____ Exp. Date _____ Signature _____

Sponsor's Name/Who introduced you to AWHP? *(Optional)* _____

You will begin receiving services upon receipt of payment. Please allow 4-6 weeks for initial receipt of publications.

If you have any questions regarding your membership services, please call AWHP at (847) 480-9574, fax (847) 480-9282.
Mail completed application to: AWHP, 60 Revere Dr., Ste. 500, Northbrook, IL 60062